World Meliponine Taxonomy Nomenclature

Abu Hassan Jalil

IBRA
INTERNATIONAL BEE RESEARCH ASSOCIATION

Jointly published by: The International Bee Research Association,
a Company Limited by Guarantee, 1, Agincourt Street, Monmouth, NP25 3DZ (UK) &
Northern Bee Books, Scout Bottom Farm, Mytholmroyd, Hebden Bridge HX7 SJS (UK).

Obtainable from:
www.ibra.org.uk & www.northernbeebooks.co.uk

ISBN: 978-1-913811-04-4

Cover design and artwork by DM Design & Print
IBRA Proof Editor - Stuart A. Roberts

Contents

Editor's Notes

by Dr. David Ward Roubik

Photograph description

Img 1 "Here's a recent portrait, in my office- where I have a very large nest entrance of Trigona in my hand, and a traditional hive of Melipona beecheii on the desk." Dr David Ward Roubik Photo by Eric Tourneret

The very large nest entrance I hold in the picture was found in Peru, at the Los Amigos field station, in Madre de Dios, where I filmed "Deep Jungles" with a film crew from the BBC in 2005. Only a part of the nest entrance is shown- its tube was about 80 cm long, making it the only stingless bee, besides the Asian *Geniotrigona*, to build such a long (and heavy) nest entrance tube. The *M. beecheii* hive is a gift from INNI (Indigenous Affairs Office of the Maya) at Chan Santa Cruz, Quintana Roo, Mexico. It is over 100 years old, is normally blocked by a single chunk of stone at each end and is of the Sapotaceae *Manilkara*- one and the same tree that chicle, or chewing gum, the sap is taken from. If you think that these details or the picture are useful for your book, then it is fine by me to use them. Those huge bees, in this case, are all orchid bees- one of my favourite subjects- the big green ones are parasites in the nests of the big striped ones. Img 1

Comments on manuscript

General comments are; watch out for italics and spelling- Plebeia is misspelt a few times, and italics lacks both plant and animal names – [scientific names are italicized because they are in Latin. When writing in English, words or phrases in another language should be italicized to help the reader know they're not English and may be subject to different pronunciation rules.]

A suggestion is that the term TOPOTYPE might need an introduction, and also LECTOTYPE and NEOTYPE.

Topotypes are about all I want right now, after seeing the types. A topotype is usually a much better specimen, can be dissected or analyzed chemically, and is not limited to the actual type series or holotype. It is often needed to make adequate microscopic photos- like some that I have included herein.

Unpublished names

You did mention something in your letter that I think is urgent to 'confront'.

When unpublished names are published, that makes them nomina nuda or a pseudo-synonym, and they have to be incorporated in any taxonomic works in the future- which is not very convenient. Mostly, as I see it, aside from bad entomological practice, it also makes one worry/wonder about type specimens and what they represent- and most' manuscript names' do not have designated types, so it is REALLY confusing. To me, and I've heard this often from Dr. Michener, it seems best to leave out the names of unpublished taxa. They could be designated by other things- like the describing author's initials and a number or letter. For instance, RLD-14, or DWRTrigoniscaC. Also, for determination labels, the year is an important character to include- syntheses or new studies often come along, and then the taxonomy also changes for those who keep up.

Now that I seem to have fallen into a never-ending spiral of new species of bees to curate and describe, the time flies. I have about 32 meliponines from the Neotropics, with no names, and eight euglossines, and that is without looking too hard. This seems to be the priority for the near future

This book is going to be pondered and consulted by many students of the great tribe of the ancient stingless honeybees. Please continue.

Yours truly,
David Ward Roubik
Senior Scientist
Smithsonian Tropical Research Institute
Republic of Panama

Foreword by Dr. Claus Rasmussen

Stingless bees – a world of diversity – also in their names

Img 2 nr. Huallaga river between Juan Guerra and Picota, San Martin, Peru. A bee nest spotted above road...

Stingless bees were historically an important source of honey and wax in tropical and subtropical parts of the world. From ancient cultures in Central America to Africa and Asia, the bees were always appraised for their small quantities of honey and wax. In India and neighbouring regions, the bees were known as 'dammar' bees in reference to the resin gathered by the bees and extracted from their hives for use, e.g., in waterproofing wooden ships. Recently stingless bees are again gaining popularity. The renewed interest is prompted by an interest in natural health products, such as their honey, but also due to the ecosystem service, the pollination, that they freely contribute when they move around between flowers and securing better seed-set and fruits.

There are hundreds of different species of stingless bees with a myriad of scientific names attached. These names can cause confusion and problems when memorizing or when attempting to communicate about the different species. The closely related honey bee (*Apis*) is only represented by a single species in many parts of the world (*Apis mellifera*). In Southeast Asia, there are several species of *Apis*, but rarely more than three in any single area. Thus, with a couple of names, it is easy to communicate about different species of honey bees in any part of the world. This is much more complicated when it comes to stingless bees. A single location can host up to 50 different species (Rasmussen & Gonzalez 2009), each one with a distinct scientific name but rarely a common name. Working globally and keeping track of hundreds of names becomes even more complex, as scientific names can be derived from names of localities, remarkable features of the bees or names of scientists. The meaning of the name is often long forgotten or was never explicitly stated by the person proposing the new name. Therefore, it is great to help that Abu Hassan has compiled and deciphered the meaning behind each of the names proposed for stingless bees. This valuable resource can provide us with a better tool for remembering names for spelling names correctly and provide a window into the world of naming new species. Interestingly, when naming the new taxa of stingless bees, the most often used inspiration is derived from the bees' morphology or colouration. This is followed in popularity by names proposed in honour of other people and then locality names.

The book opens with a general introduction to the scientific naming process and explains why such names can change over time. This is followed by a section of Meliponine taxa definitions where the author, in a hierarchical form, explains the meaning of all of the genus-group or higher-level names; this includes the names for all genera, the tribe, subfamily, family, and order Hymenoptera. The following section on patronymics lists all persons appearing as part of the stingless bees' scientific names. This section is actually longer than one would expect. About 178 times have new names been proposed for stingless bees in honour of a person, including many scientists, collectors, naturalists, friends and even the Princess of Thailand. The next section is an exercise in geography tracing the geographical origin of about 175 different names derived from either the collecting locality or the distribution range from where they are known. The fascinating life of stingless bees is hinted at in the section on floral and faunal interconnections. The author has identified 33 names that were derived by some biological interaction. Many, of course, from the floral hosts of the bees and other insects and animals associated with the bee nests have been used as inspiration while naming new stingless bee species. This is followed by more than 300 descriptive names derived from the morphology or anatomy of the stingless bee species. This can include anything from the colour of the wing to the shape of the legs. The last section on the names includes those that could not be placed in any of the above sections. The compilation neatly closes with an index to the meaning of the general epithets for stingless bee taxa, such as the root words used when constructing the scientific names, e.g., the meaning of 'cephala' and others. The glossary and index make it an easy to use tool, and I am excited to introduce this new book on stingless bees and their diverse names, providing an important historical account and a unique compilation of information related to their names and the naming of the bees. The author of this book has done a terrific job putting this together, and I hope that stingless bee enthusiasts will find it stimulating and helpful in tracing the names of the stingless bees.

Claus Rasmussen, PhD.

Literature cited:

Rasmussen, C. & Gonzalez, V.H. (2009) Abejas sin aguijón del Cerro Escalera, San Martín, Perú (Hymenoptera: Apidae: Meliponini). *Sistemas Agroecológicos y Modelos Biomatemáticos (Tarapoto)*, 2, 26–32.

Foreword by Dr. Anne Dollin

Img 3 After exploring the stingless bees in the most remote parts of Australia for 30 years with my husband, Les, I am now analysing our collection for an Austroplebeia revision paper. Here I am visiting London's Natural History Museum, examining 170 year old type specimens from far north Australia as part of our revision work. (photograph taken by Les Dollin)

Interest in keeping stingless bees is booming. These safe and easy-to-manage bees are ideal for schools, community gardens and enthusiasts; their honey and propolis are revealing beneficial medicinal properties, and they are proving to be valuable pollinators for agricultural crops.

Unfortunately, the dozens of species of stingless bees in the Indo-Malayan and Australasian region can be difficult to distinguish and have few common names. Australian *Tetragonula hockingsi* and *T. carbonaria*, for instance, can only be reliably distinguished by their different nest structures or their DNA. A confident knowledge of their scientific names is vital to understanding these intriguing species.

This new book by Abu Hassan Jalil explains the scientific names of the many stingless bee species in our region and will be a valuable tool for enthusiasts and researchers. Most scientific names are Latin or Greek versions of the names of people, places or characteristics. Knowing the meaning of those names can help us spell them and remember them. Moreover, the names often give fascinating glimpses into these species' history, often long forgotten.

The species names covered in Mr Jalil's book are divided into five sections.

First are the bees named in honour of people, such as influential scientists, collectors and naturalists. For instance, *T. biroi* was named after the outstanding Hungarian collector, Lajos Biro, who explored the insects of New Guinea from 1896 to 1901, finding thousands of new species. *T. hockingsi* was named after the early Australian naturalist Harold Hockings, who wrote one of the first accounts of our stingless bees, published in 1884.

Next are the bee species named after ethnic groups and localities. *Austroplebeia essingtoni*, for instance, was named after Port Essington, the site of one of the earliest British settlements in remote far north Australia. The settlement failed after only eleven years due to disease and harsh conditions. However, two tiny stingless bees, collected there and sent to England on a sailing ship in 1842, became the type specimens of a major Australian species.

The bees in the next section were named after associated flora or fauna. For instance, *T. angophorae* was named after the *Angophora* gum trees that attract clouds of stingless bees when they flower.

The fourth and largest section lists species named after structural features or colours of the bees themselves. Some names have clear meanings, such as *A. cincta,* named after the bright yellow band on its thorax, and *T. iridipennis,* named after its iridescent wings. Others are puzzling, such as the Australian black bee, *T. mellipes*, named after its 'honey-water coloured feet'! Later investigations by Professor Shôichi F. Sakagami (Dollin et al. 1997) revealed that the type specimen of this species was probably a pale-coloured immature bee from inside the nest.

Finally, there is a list of other names, often relating to behaviour or general appearance. For instance, the name of the minute stingless bee, *T. atomella*, means tiny, like an atom.

Mr Jalil's book concludes with a reference list of root words commonly used in these scientific names and a glossary of important taxonomic terms. We congratulate Mr Jalil on this extensively researched book. It will shed great light on the often-mysterious scientific names of the species in our region and help us all better understand our fascinating native stingless bees.

Dr Anne Dollin

Australian Native Bee Research Centre

and Aussie Bee Website www.aussiebee.com.au

****

Reference cited in Foreword:

Dollin, A.E., Dollin, L.J., & Sakagami, S.F. (1997) Australian stingless bees of the genus *Trigona* (Hymenoptera: Apidae). *Invertebrate Taxonomy,* 11, 861–896.

Acknowledgements

I am indebted to Dr. Ibrahim bin Shuib, a Medical Practitioner who introduced me to the culture of Stingless Bee Keeping and subsequently the name Trigona. It was his mentorship that urged me to develop this etymological compendium for the Stingless bees.

I also extend my gratitude to Dr. Claus Rasmussen, a well-known corbiculate bee biologist who accorded me many leads and suggestions in the making of this compilation. He also contributed many close-up photographs of Peruvian meliponines and microscopic images of Indian stingless bees. I am also grateful for Dr. Rasmussen's valuable notes and advice without which, this compendium would not have been possible.

Much appreciation is due for Dr. Anne Dollin of the Australian Native Bee Research Centre (ANBRC), who so unselfishly contributed a couple of photographs of Austroplebeia cincta. She also did help fine-tune the entries for the Australasian clade.

Thanks to Dr. Ricardo Ayala, who allowed me to use his Mexican Stingless Bees' excellent photographs.

Permission to use photographs of many New World bees were from Dr. John S. Ascher and Dr. Giorgio Venturieri. Argentina stingless bees images were contributed by Maria Rosa Bernasconi and from Australia, contributed by Robert Luttrell, while many from Brazil were from Luiz Carlos da Cunha.

Special thanks go to Prof Dr. Peter Kwapong for his contribution of some African stingless bees images and to Dr. David W. Roubik for contributing some images and for taking the trouble to look through and helped with the final editing of this lengthy manuscript.

I also thank the many individual macro photographers who contributed and are listed in the List of images on pages 231 -234.

Inspirations

A deserving mention is Dr. Hans Bänziger, from the Chiang Mai University of Thailand, who inspired me with his short linguistic biography. This, he related to me when I told him that I was writing these notes on the names of biological taxa.

Img 4 Dr. Hans Bänziger, from Chiang Mai University of Thailand

"As an Italian-born Swiss, I grew up multilingual, a daily struggle because I spoke Swiss dialect at home, outside it Italian, and at school German. Our first foreign language was French, the second English. At school, I mostly was just good enough to be mediocre to pretty bad in languages. Today English is my everyday scientific language, and it is still sorts of a struggle. My daily language for the past decades is Thai, my fifth (sixth?) language, both at home and at the university. Being polyglot may be an advantage in certain respects (though certainly not when one has to write a language correctly), but it does not come without an effort."

'This enterprise at the limestone face, six times over a period of four years, will remain firmly impressed in my mind. The site required a strenuous, steep and slippery three-hour hike to reach, so we bivouacked there up to three consecutive nights. We not only discovered there the beautiful 'coral'-like nest entrance of the stingless bee Pariotrigona klossi. We were also successful in the study of the pollination of the rare lady slipper orchid Paphiopedilum thaianum, which actually was the original scope of our enterprise. This beauty is pollinated by the halictid bee *Lasioglossum orchidodeceptum*, then a new species. We baptized the bee as '*orchidodeceptum*' because it is deceived by the orchid to pollinate it without receiving nectar as a reward. Finally, to top it all, I had to perform a feat I had never done before: I had to climb the rock face by night's darkness in order to check whether these stingless bees close the entrance of their nest at night. They don't.

Hans Bänziger & Sängdao Bänziger are the authors of "Mammals, birds and reptiles as hosts of *Lisotrigona* bees, the tear drinkers with the broadest host range (Hymenoptera, Apidae)" *BULLETIN DE LA SOCIÉTÉ ENTOMOLOGIQUE SUISSE* 83: 271–282, 2010

The left photo is of *Lisotrigona cacciae* Img 6, the one on the right is of *Pariotrigona klossi* Img 5 drinking tears from Dr. Hans Bänziger own eyes.

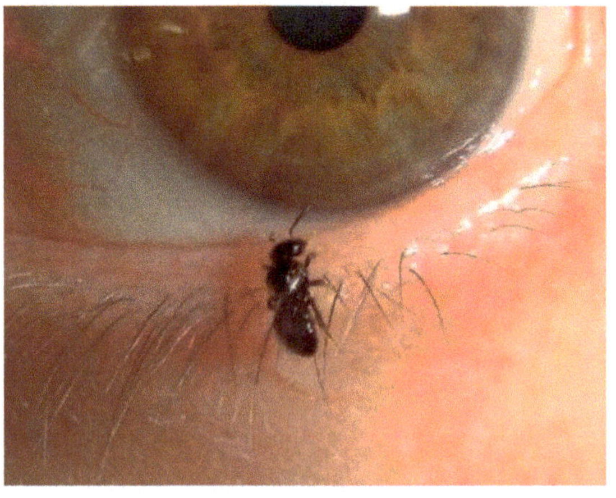

Img 6 *Lisotrigona cacciae*

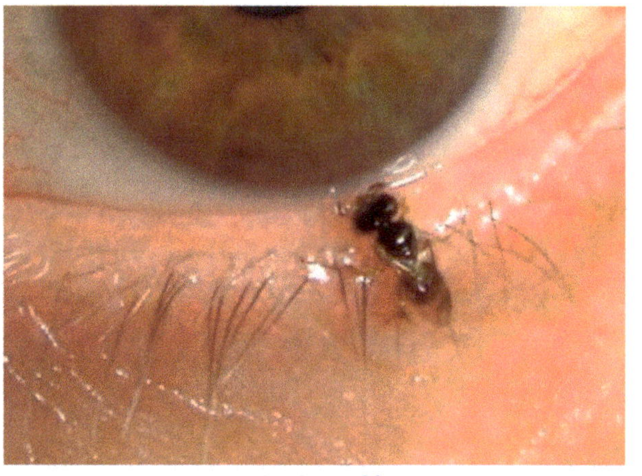

Img 5 Pariotrigona klossi

Preface

Many have wondered what it was like to discover new species and the subsequent processes that verify and ultimately name such a new species. Appropriately, for readers who desire a glimpse of how and why stingless bees are named the way they are, this book gives an inkling of the complex and timely process of naming and systematics. The stingless bees name from as long ago as 1806 (by Johann Karl Wilhelm Illiger, a German entomologist and zoologist) (Wikipedia, Johann Illiger, 2014) as in *Melipona* (Illiger 1806) and also 1807 (by a Swiss Physician, Louis Jurine, see Portrait 1) (Wikipedia, Louis Jurine, 2014) with *Trigona* (Jurine 1807) are a testament to the Linnean Taxonomy (Biological Classification of Living things) of 1735 with the naming of animals in the 10th edition in 1758. Biological classification, or scientific classification in biology, is a method of scientific taxonomy used to group and categorize organisms into groups in a hierarchical way, such as genus or species. These groups are known as taxa (singular: taxon).

Portrait 1 (Wikipedia, Louis Jurine, 2014)

In order to fully understand biological evolution, we need to be aware of how animals and insects interact and coexist in the animal kingdom. Comparing and relating the evolutionary variations among different species is important to this understanding. Besides, it is interesting to learn about other kinds of creatures, especially insects.

- "King Penguins Congregate on Frozen Ground Sometimes"
- "King Parrots Clash Over Fairly Good Seed"
- "Keep Pond Clean or Frogs Get Sick"

These sentences make little sense to a layperson, but to a zoology student, they are mnemonics used to memorize the scientific classification applied in zoology. When applied, it refers to the sequence of "Kingdom, Phyla, Class, Order, Family, Genus, Species". This book will help readers memorize the many different names of the varied species of stingless bees from throughout the tropical world.

This book also provides synonymy were available to lend descriptive insights to the species anatomical attributes, colour attributes and also to certain patronymics, although they may not help in species recognition but at least in recognition of the prominent people. The names used to depict locations help in the geographical availability, and names that reflect interactions with specific flora and fauna lends to its habitat and regions. However, these names do not help actual recognition but bring some light when listed with their synonyms. The synonyms are gathered from 3 main catalogues, namely The New World Bee Catalogue (Moure's Bee Catalogue, 2008), Catalogue of Afrotropical Bees (Eardley & Urban, 2010), and Catalog of the Indo-Malayan/Australasian stingless bees (Rasmussen, 2008) and besides, for Madagascar stingless bees based on Combining morphology, and DNA barcoding resolves taxonomy of Western Malagasy *Liotrigona* Moure, 1961 by Hauke Koch (*#Koch, 2010).

The compilation includes data and names up to mid-2016 and possibly has included all names ever used in naming Meliponines The information was gathered from numerous papers, books, bulletins, journals, historical data as far back as 1836 (e,g, Lepeletier de Saint Fargeau, A.L.M. (1836). Histoire Naturelle des Insectes) and electronic media in the worldwide web. Although the compilation took an extensive length of time and a varied source of information, details may have slipped from this compilation. Any errors and omissions may correspond to abuhsn@gmail.com.

Words that inspired this work.

“Understanding our past determines actively our ability to understand the present.”

(Anonymous)

Patience abide; God does provide.

(Theodora Johansen 1921-1985)

Chapter 1

Introduction

Etymology

Though it is well-known that taxonomy owes its roots to Latin and Greek, one cannot help but wonder where peculiar words and terms were derived from. In the study of Meliponiculture (beekeeping with stingless bees), one may encounter species *like Tetragonula* ***sirindhornae*** (Michener & Boongird and *Lepidotrigona* ***hoozana*** (Strand) from the Indo-Malayan region or *Lestrimelitta* ***nitkib***, *Celetrigona* ***manauara***, *Dolichotrigona* ***chachapoya*** from the Americas, or *Liotrigona* ***madecassa***, *L.* ***bitika***, *L.* ***mahafalya***, from Madagascar. These are just a few species names that are neither Latin nor Greek in origin. Another good example is *Heterotrigona* ***itama*** from proto-Malay **itam,** meaning dark or devoid of light.

Then we have examples like *Trigona* ***amalthea*** and *Trigona* ***zebra,*** which are epithets. An **epithet** (from Greek: ἐπίθετον *epitheton*, neut. of ἐπίθετος *epithetos*, "attributed, added") or **byname** is a descriptive term (word or phrase) accompanying or occurring in place of a name and having entered common usage. It can be described as a glorified nickname.

Hence, in that search to find the meaning or why or how names are derived, one arrives at **Etymology**. Etymology means studying the history of words, their origins, and how their form and meaning have changed over time. By extension, the term "the etymology of [a word]" means the particular word's origin.

Meliponine species named after locations or regions(e.g. *Tetragonula* ***sarawakensis*** or *T.* ***minangkabau*** in the Indo-Malayan region, *Melipona* ***belizeana,*** *M.* ***boliviana*** or *M.* ***costaricaensis*** from the Americas and *Axestotrigona* ***cameroonensis*** or *Meliplebeia* ***tanganyikae***) where they are found/collected play a very important role as insect evidence in forensic entomology. Identification of the species can tell where they are from, especially in stingless bees because they are non-migratory(or at least limited migration due to small flight range in that their geographical distribution is quite regionally concentrated).

Some were even named after the botanical species the bees were discovered upon, e.g. *Cephalotrigona* ***zexmeniae*** (Cockerell), presumably for Zexmania (Asteraceae), *Melipona* *fulvipes* ***triplaridis*** Cockerell, for *Triplaris cumingiana* in Panama and *Melipona schausi* ***convolvuli*** Cockerell for Convolvulus (the Morning Glory family) in Venezuela. Others may pay tribute to the collector, e.g. *Melipona* ***schausi*** Cockerell, where the type specimen indicates that it was from a collection made by William Schaus in French Guiana.

On certain occasions, one may encounter the name of the collectors' wife (who may be Entomologists themselves or specimen collectors in their own right) being honoured where a clear example is *Plebeia* ***margaritae*** Moure, *Etymology:* homage to Col. M. Alvarenga's wife **Margarita**. Another example is *Melipona* ***lupitae*** Ayala, where *Etymology shows:* homage to **Lupita**, wife of R. **Ayala**.

Synonymy

While sifting through the various Etymology, I find that some names are synonyms. In scientific terminology, **synonyms** are different scientific names used for a single taxon.

In zoological nomenclature, synonyms are different scientific names that pertain to the same taxon, for example, two names for the same species. The zoological nomenclature rule is that the first name to be published is the **senior synonym**; others are **junior synonyms** and should not be used. However, suppose the earliest name cannot be used (for example, because the same spelling had previously been used for a name established for another taxon), then the next available junior synonym must be used for the taxon.

Synonyms are "objective" if they unambiguously refer to the same taxon; this is the case if they refer to the same description or type of specimen. Otherwise, the synonyms are "subjective", meaning there is room for debate: one researcher might consider the two names to refer to the same taxon, another might disagree.

Some synonyms and also some Nomen nodum, especially the Brazilian meliponines, were adapted from folk names that take root from the Tupi Language, e.g. coeirupu (**coyrepu**) as in *Melipona* ***coyrepu*** Müller, 1875, with jatahy (**jatiformis**) as in *Trigona* Cockerell 1912, or **mandacaia** for *Melipona* ***mandacaia*** Smith, 1863, and **mombuca** for *Trigona* ***mombuca*** Smith, 1863. **Old Tupi** or **Classical Tupi** is an extinct Tupian language spoken by the native Tupi of Brazil, mostly those who lived close to the sea. It belongs to the Tupi–Guarani language family and has a written history spanning the 16th, 17th and early 18th centuries (see appendix for some Tupi-Guarani terms).

Homonym

Besides synonyms, there are also homonyms. The word **homonym** comes from the Greek ὁμώνυμος (*homonumos*), meaning "having the same name", which is the conjunction of ὁμός (*homos*), "common, same" and ὄνομα (*onoma*) meaning "name". Thus, it refers to two or more distinct concepts sharing the "same name" or signifier, {e.g. ***Trigona longipes*** Smith, 1863 (junior **homonym** of ***Trigona longipes*** Smith, 1854) spurring a new name for the earlier (1863) one. Thus ***Melipona longicrus*** Dalla Torre, (1896) (**nomen novum** for ***Trigona longipes*** Smith, 1863, non-Smith, 1854) .}

A homonym worth mentioning here is ***Trigona thoracica*** Cresson, (1878) Descriptions of new species of North American bees. *Proc. Acad. Nat. Sci. Phila.* **30:** 181-182 [181] (junior homonym of ***Trigona thoracica*** Smith, 1857, from Singapore, Malaysia.). *Type data:* syntype **ANSP** 2176 <F> worker. *Type locality:* Mexico, Sumichrast, F. This clearly shows a single name referring to two different species from different locales.

Latin Terms

The phrase **nomen nudum** (plural **nomina nuda**) is a Latin term, meaning "naked name", used in taxonomy (especially in zoological and botanical). It may or may not be written in italics, depending on style.

The term is used to indicate a designation that looks exactly like a scientific name of an organism and may well have originally been intended to be a scientific name, but fails to be one because it has not (or has not yet) been published with an adequate description (or a reference to such a description), and thus is a "bare" or "naked" name, one which cannot be accepted as it currently stands.

Because a *nomen nudum* fails to qualify as a formal scientific name, a later author can publish a real scientific name identical in spelling. If one and the same author puts a name in print, first as a *nomen nudum* and later on publishes it for real, accompanied by a description that does meet the formal requirements, then the date of publication of the latter, formally correct publication becomes the name's date of establishment.

Then there is taxon which is considered incertae sedis. ***Incertae sedis*** (Latin for "of uncertain placement") is a term used to define a taxonomic group whose broader relationships are unknown or undefined. Alternatively, such groups are frequently referred to as "enigmatic taxa". Uncertainty at specific taxonomic levels is attributed by *incertae familiae* (of the uncertain family), *incerti subordinis* (of uncertain suborder), *incerti ordinis* (of uncertain order) and similar terms.

For this reason, I consider all names valid, synonym or nomen nudum, in this compilation of Etymology of Meliponine Taxonomy nomenclature.

How Species Are Described

There has been a considerable discussion in the literature about the difference between taxonomy and systematics–which term was used first, which is most comprehensive, and so on–and some practitioners still say there is no distinction between the terms. However, one generally accepted definition is that *systematics* study of biological diversity and the evolutionary relationships among organisms (Simpson 1961; Mayr 1969; Wilson 1985). By this definition, *taxonomy* is a subdivision of systematics, consisting of three associated activities: *identification* (referring a specimen to a previously classified and named group), *classification* (ordering organisms into groups based on perceived similarities or differences), and *nomenclature* (naming groups of organisms according to rules developed for the process). Systematics also includes the study of the process of evolution and phylogeny. According to the current biological nomenclature rules, the taxonomic procedure is the practical process of identifying, recognizing, researching or redescribing a taxon for scientific publication. That being said, uniquely, an example of "evolutionary relationships among organisms" regarding meliponines is the naming of *Trigonisca* ***ceophloei*** (Schwarz, 1938) based on *Type locality:* Kartabo, British Guiana, 28.VII.1920, from the stomach of a woodpecker, ***Ceophloeus*** *lineatus* (Aves, Picidae).

Biological nomenclature

*T*he scientific naming of organisms was developed to ensure that every organism can have an unambiguous name (refers only to that particular kind of organism) and be globally understood. A species becomes known in the scientific sense when a Latin *binomial,* a name consisting of two parts (a genus term and a species term), and a description are published in the scientific literature, according to the rules of botanical (The International Code of Botanical Nomenclature, ICBN) or zoological (The International Code of Zoological Nomenclature, ICZN) nomenclature. Although they are written in legal language, there is no agency to enforce these codes; only biologists agree to observe and accept them (Jeffrey 1989), [Winston 1999].

Meliponini

Stingless bees, sometimes called **stingless honey bees** or simply **meliponines**, are a large group of bees (approximately 500 species), comprising the tribe **Meliponini** (or subtribe Meliponina according to other authors). They belong in the family Apidae and are closely related to common honey bees, carpenter, orchid bees and bumblebees. The common name is slightly misleading as male bees and bees of other species, such as Andrenidae, cannot sting. Meliponines have stingers, but they are highly reduced and cannot be used for defence.

Fig 1 Dominican Amber Specimen: HYMENOPTERA Apidae: Meliponini Proplebeia dominicana (Stingless Bee) Possibly 23mya (typical of Dominican Amber)

The extinct genus *Proplebeia* **Error! Reference source not found.**are recognized four species, three of them from the amber of the Dominican Republic and one from the amber of Chiapas, Mexico, dating from the Miocene (see Camargo *et al.*, 2000). ***Cretorigona prisca*** has until recently been considered the oldest fossil bee. In 2006 another oldest fossil Hymenoptera, *Melittosphex burmensis* was found.

To better comprehend the taxon used in naming stingless bees, I find it imperative to compile a list of the derivative nomenclature, including prefixes, suffixes and root words that comprise the naming structure. The reader will recognize the geographical locations and names of Naturalists involved in the study of bees in general because many species were named after these researchers and locations of where they were collected or observed.

Other terminology, basically Latin, describes the anatomy characteristics and/or colour. This allows easier identification by simply identifying location observed/recorded, anatomy description and colour. Preferably used with an entomological glossary. These seemingly exhaustive definitions are not certified specifics of the taxonomical definitions, but merely a listing of possible lingual explanations of the reasons for naming the nomenclature was structured. It is intended for the layman beekeeper and enthusiast to recognize the bees' anatomical description and characteristics from their indicative names. The names of the naturalists, researchers and collectors are also honoured accordingly in the validated taxa.

Trigona

tri·go·na (trī-gō'nă) = Plural of trigonum. Definition of *TRIGONUM*: a triangular anatomical part: Triangular Anatomical features of the stingless bee (Fig. 1 & Fig. 2)

1) **cephalic triangle** on cranium formed by lines connecting the metopion, the pogonion, and the occipital point.
2) **clypeal triangle,** the fronto-clypeal suture forms a 'supra-clypeal **triangle.**'
3) **digastric triangle** submandibular **triangle**.
4) **facial triangle** The area that is raised and extends from between the toruli to the posterior margin of the clypeus
5) **occipital triangle**. opening where organs of head and body are connected
6) **ocellar triangle** typically three bead-like lenses arranged in a **triangle**
7) **propodeal** triangle of propodeum
8) **tibial triangle** triangularly clavate hind tibia
9) **tarsal triangle** triangular shape of tarsus

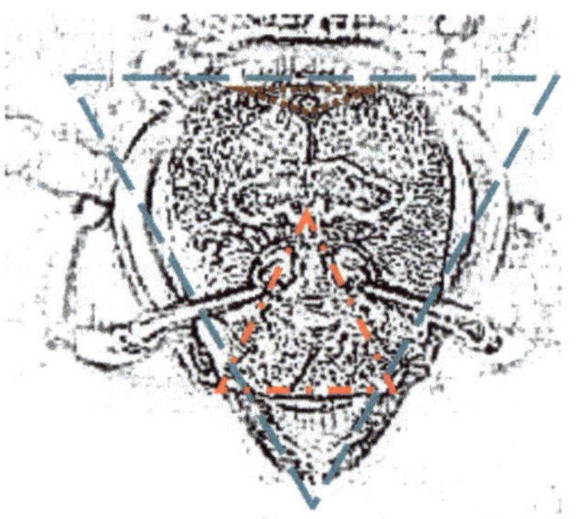
Fig. 1 Front view of Triangular Head

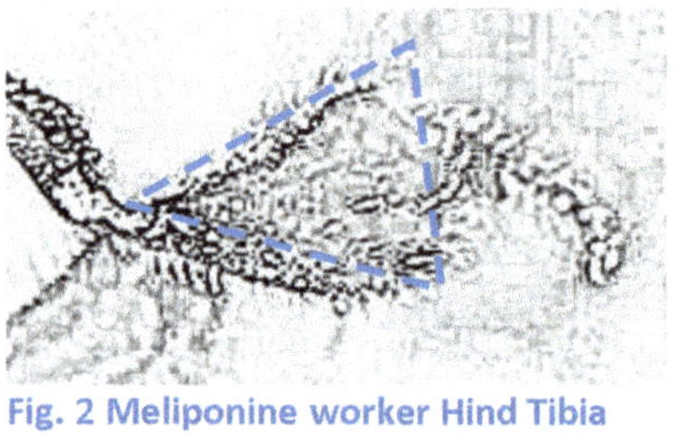
Fig. 2 Meliponine worker Hind Tibia

Binomial Nomenclature

In biology, the **binomial** or **binomial** (also called **binary nomenclature** or **binary name**) is a standard convention for naming the different species of organisms (living or extinct). Sometimes referred to as binomial nomenclature *binomial classification system*.

As suggested by the word "binomial", the scientific name assigned to a species is formed by the combination of two words ("names" in Latin or Greco-Roman root): the genus name and specific epithet. The combination of both is the scientific name that identifies each species as having "full name".

Value and use

In the scientific context, the usefulness of the binary formula is only to save the ambiguity that can occur with different common names for an organism and to name those specimens that do not even have a common name. It also overcomes the difficulties of communicating in different languages from the universal recognition and agreed standard naming system.

However, despite the rules that dictate the binary name uniqueness for a species, in practice, it is common that there are synonyms; that is, there are several outstanding scientific names for the same species (generally dependent point of view of a particular taxonomic system in use, and ultimately, the author).

The names' stability is sometimes compromised by the resurrection of some names forgotten in time, claiming its priority to be first published. However, in these cases, it is possible to keep these names (*nomina conservanda* or *nom. cons.*) according to nomenclature codes in use. For zoological nomenclature, are accepted as valid only those names from 1758 appear in an official publication.

Trinomial nomenclature

The trinomial nomenclature (or trinomial name) is an extended form of the binomial used in biology to identify each species of living being known to refer specifically to smaller subdivisions of a given species. The most common of these are:

- The subspecies, in which case a third name is always added tiny behind those related to gender (capitalized and first) and species (lowercase and second).

- The variety or race, very little definite to be considered a subspecies, identified then typing the name of the species or subspecies the abbreviation "var." followed by the name of the variety itself.

Less used but also outside the trinomial nomenclature are the submanifold ("subvar."), Form ("form" or "f.")

History

Adopting a system based on two names nomenclature is due to Swedish physician and naturalist Carl Linnaeus (in Latin, Carolus Linnaeus, 1707-1778), who attempted to describe the entire known natural world giving "every kind" a name consisting of two parts. However, the binomial nomenclature Linnaeus existed before in various forms. But from Linnaean publications begins to generalize the use of binomial nomenclature for specific.

Nomenclature codes

Since the mid-nineteenth century, it became increasingly apparent the need for a body of rules which shall regulate the establishment of scientific names. Over time, these canons, known as *Nomenclature Codes*, dictated the name of:

- animals (International Code of Zoological Nomenclature),
- plants (International Code of Botanical Nomenclature), including fungi and cyanobacteria,
- bacteria (International Code of Nomenclature of Bacteria), and
- virus (International Code of Nomenclature of Virus).

The rules of nomenclature codes vary from one to another.

Aspects of Binary Nomenclature

There are some aspects universally adopted for the formulation of binary nomenclature.

The person who first described a species (the "author") is the one that has the privilege of naming. When the author chooses to name or specific epithet derived from a proper name, it is a title of honour or recognition, dedicating it to a colleague, friend or family, in Latinized form, is considered rude and a sign of selfishness the author devotes himself species. In this book, this is classified under "**Patronymics & Matronymics**".

In other cases, the names are chosen in reference to the character or properties of living that is named. Henceforth, this is classified in sections "**Peoples & Places**", **"Floral & Faunal Relationship"**, and **"Anatomy & Colour Attributes"**, with the other epithets in the general epithets section.

Chapter 2

Meliponine Taxa Definitions

Order

Order - Hymenoptera (Greek, ***hymen*** = membrane; ***pteron*** = wing)

Description

Hymenoptera is easily recognized as belonging to the order because of their "wasp waists". The constricted waist (actually a constriction in the front part of the abdomen) characterizes ants, wasps, parasitic and bees in the suborder Apocrita. They vary from minute parasitoids of other insect eggs up to huge wasps (up to 2 inches in size) and bees, but all have the wasp waist(with exceptions of Symphyta. The "waist," or midsection, between the abdomen and thorax, is broad in Symphyta)

Superfamily

The superfamily **Apoidea** is a major group within the Hymenoptera, which includes two traditionally recognized lineages, the "sphecoid" wasps and the **bees.**

Nomenclature

Bees appear in recent classifications to be a specialized lineage of crabronid wasps that switched to pollen and nectar as larval food rather than insect prey; this presumably makes the Crabronidae a paraphyletic group.

As bees (not including their wasp ancestors) are still considered a monophyletic group, it is still convenient to use a grouping between superfamily and family to unify all bees. A few recent classifications have addressed this problem by lumping all bee families together into the single large family Apidae, though this has not met widespread acceptance. The alternative classification in more common use is to unite all bees under the name Anthophila (Engel, 2005), which is equivalent to the obsolete name Apiformes (which meant 'bee-like forms' in Latin)

Family/Subfamily

Apidae is a large family of bees, comprising the common honey bees, **stingless bees** (also used for honey production), carpenter, orchid bees, cuckoo, bumblebees, and various other less well-known tribes and groups. Many are valuable pollinators in natural habitats and for crops.

The **Apinae** is the subfamily that includes the majority of bees in the family Apidae. It includes the familiar "corbiculate" (pollen basket) bees —honey bees, **stingless bees**, orchid bees, and bumblebees.

Meliponinae = Melipona [Latin **Meli** (honey) + **pona** (Latin "I make")] + -inae [Translingual Suffix **–inae-** (*taxonomy*) *Used to form names of subfamily of animals*]

Tribe Taxon: Root words

This section suggests Root words of the terms used in describing the family and tribe taxon. Stingless Bees are of the Family Apidae – Long Tongued Bees.

Melipona = Latin **Meli** (honey) + **pona** (Latin "I make")

Meliponini - A taxonomic tribe within the subfamily Apinae, Meliponini = Translingual – Etymology: - *Melipona* (see Melipona) + *-ini* (see suffix –ini) –

-ini = Translingual – Etymology: New Latin, the nominative plural form of Latin -*imus* - Suffix '**–ini'** = (taxonomy) Used to form names of tribes of animals.

Meli = Borrowing from **Ancient Greek** ***μέλι*** (meli). Coined by missionaries for the 1839 translation of the Bible. From Latin **mel** n (*genitive* **mellis**); *third declension* honey (figuratively) sweetness, pleasantness

pōnī = Latin **poni -** present passive infinitive of pōnō "to be placed, to be put, to be laid."; "to be ordained."

Trigona- = From Ancient Greek τρίγωνον **(**trigōnon, "triangle"), Latin **Trigona** (used as a prefix) = **trigōnus** m (*feminine* **trigōna**, *neuter* **trigōnum**); *first/second declension* - trigonal; three-angled; having three angles. [see above diagrams list of trigona (plural of trigonum) in Stingless bees]

Trigonini = **Trigona** + **-ini** [Translingual – Etymology: New Latin, nominative plural form of Latin *-inus* - Suffix '**–ini'** = (taxonomy) Used to form names of tribes of animals.]

Note: Trigonini is no longer in use (Michener and his "Bees of the World" used only Meliponini. Michener, C.D. (2007) The bees of the world, second edition. Johns Hopkins University Press, Baltimore, xvi+953. Michener discontinued using Trigonini long before this paper: Michener, C.D. (1990) Classification of the Apidae (Hymenoptera). Appendix: Trigona genalis Friese, a hitherto unplaced New Guinea species. University of Kansas Science Bulletin, 54, 75–163.) [Claus Rasmussen stressed this in personal communication.]

Chapter 3

Meliponine Taxa: Genus/Subgenus - Epithets & Root words

This section suggests the prefixes and meaning of root words used in describing the meliponine Genus Taxon. Where the derivative word is from Ancient Greek, the spelling in Greek alphabets is shown.

Suffix **-trigona** = **From Ancient Greek** τρίγωνον **(trigōnon, "triangle"),** Latin **Trigona** = **trigōnus** m (*feminine* **trigōna**, *neuter* **trigōnum**); *first/second declension* - trigonal; three-angled; having three angles. [see above diagrams list of trigona (plural of trigonum) in Stingless bees]

Aparatrigona Moure, 1951* (Moure's Bee Catalogue, 2008) = From Latin *apparāre*, present active infinitive of *apparō* ("to prepare"). (transitive) to protect, defend, guard + trigona

Apis Linnaeus, 1758* (Moure's Bee Catalogue, 2008) = Latin **apis** f (*genitive* **apis**); *third declension* - A bee.

Apotrigona Moure, 1961^ (Eardley, 2004) = Latin-*present active* **apō**, *present infinitive* **apere**, *perfect active* **apī**, *supine* **aptum** - attach, connect, join, bind + trigona

Austroplebeia Moure, 1961# (Rasmussen, 2008) = Austro refers to Australia as a suffix in the genus *Austroplebeia* - a Plebeia bee in Australia. (Rasmussen, Personal communication, 2013)

Axestotrigona Moure, 1961^ (Eardley, 2004) = Axes (from Latin **axis**) + to (Postposition **–to**) + trigona

Genus 1 *Cephalotrigona* sp. (Mombucão) Photo: Claus Rasmussen

Celetrigona Moure, 1950 = Spanish **cele** (*infinitive* **celar**) - First-person singular (*yo*) present subjunctive form of celar. - to watch over, to conceal + trigona (first introduced as a subgenus of *Hypotrigona* Cockerell, 1934) (Moure's Bee Catalogue, 2008) .

Camargoia Moure, 1989 = [homage to Prof. João Maria Franco de Camargo.]* (Moure's Bee Catalogue, 2008)

Genus 2 *Geniotrigona* sp. (Kelulut Besar) Photo: Ahmad Zaini

Cephalotrigona Schwarz, 1940v= Latin prefix in (biology) relating to the brain or head + trigona (first introduced as a subgenus of *Trigona* Jurine, 1807). (Moure's Bee Catalogue, 2008)**Genus 1**

Genus 3 *Geotrigona* sp. Photo: via Maria Rosa bernasconi

Cleptotrigona Moure 1961 (Eardley & Urban, 2010) = from Ancient Greek **klepto-** = Forming compound words having the sense of "theft", "thief", "stealing". Used to describe stingless bees with a robbing habit. They rob pollen and nectar from other stingless bees' nests and are not known to visit flowers.

Cretotrigona Engel (ENGEL, 2001)= Stingless bees of the Late Cretaceous age

Dactylurina Cockerell 1934 (Eardley & Urban, 2010) = **dactyl-** From Ancient Greek *δάκτυλος* (daktulos, "a finger"), + **urina** - From Proto-Indo-European **ur*; compare Ancient Greek *οὖρα* ("urine"); variant of root **awer-* ("to moisten, flow"), Unlike other African stingless bees, this genus builds a vertical comb, surrounded by batumen, in exposed nests on trees (Michener, 2000).

Dolichotrigona Moure, 1950 = From Ancient Greek *δολιχός*. English Prefix **dolicho-** = long + trigona (Introduced first as a subgenus of Hypotrigona . Cockerell, 1934) (Moure's Bee Catalogue, 2008)

Duckeola Moure, 1944 = patronym for Adolf **Ducke** (Moure's Bee Catalogue, 2008) + ola (see suffix **-ola**)

Eomelipona Moure, (1992) (Moure's Bee Catalogue, 2008) = Eo (see prefix Eo-) + melipona (see Melipona)

Frieseomelitta Ihering, 1912 = honouring H. Friese (Moure's Bee Catalogue, 2008) + melitta (see suffix melitta) see also *Friesella*

Genus 4 *Friesella* sp. Photo: Wilson Gussoni

Friesella Moure, 1946 = honouring H. Friese (Moure's Bee Catalogue, 2008) + lla (see suffix –lla).; see also *Frieseomelitta*

Heinrich Friedrich August Karl Ludwig Friese (1860–1948) was an important pioneer bee biologist (melittologist). Between 1883 and 1939, he described 1,989 new species and 564 new varieties or subspecies of insects, of which over 99% were bees. (Claus Rasmussen & John S. Ascher, 2008) Genus 4

Genus 5 *Heterotrigona* sp. (Kelulut Hitam) Photo: Hisyam Ismail

Geniotrigona Moure, 1961 (Rasmussen, 2008) = Genio (prefix **English** Forms words relating to the chin + trigona (see suffix –trigona) Genus 2

Geotrigona Moure, 1943 (Introduced first as a subgenus of *Trigona* Jurine 1807.) (Moure's Bee Catalogue, 2008) = English Prefix **'geo-'about** Earth + trigona (see suffix –trigona) Genus 3

Genus 6 *Homotrigona* sp. (Kelulut Perang) Photo: Syafrizal Fachmy

Heterotrigona Schwarz, 1939 (Rasmussen, 2008)= variable in composition and colour. From Ancient Greek *ἕτερος* (heteros, "other, another, different") + trigona (**see suffix –trigona**) Genus 5

Homotrigona Moure, 1961 (Rasmussen, 2008)= Greek or Medieval Latin - Having the same composition throughout; of uniform make-up or colour. Ultimately from Ancient Greek *ὁμός* ("same") + trigona (**see suffix –trigona**) Genus 6

Hypotrigona Cockerell (1934) (Eardley & Urban, 2010) = From Ancient Greek *ὑπο-* (hypo-), combining form of *ὑπό* (hypó, "under"). Prefix (anatomy) Below; beneath; under. Less than normal. + trigona (**see suffix –trigona**)

Kelneriapis Sakagami 1978 **Kelneriapis** from Middle Eocene Baltic amber (44.1 Mya; (Kelner-Pillault, 1969, 1970); (ENGEL, 2001)

Kelnermelia Moure *In* Moure and Camargo, 1978: The generic name is in homage to Dr. S. Kelner-Pillault. [&&] (Moure J.S. & Camargo J.M.F., 1978)

Lepidotrigona Schwarz, 1939 (Rasmussen, 2008) = From Ancient Greek λεπίς (lepís, "scale") English Prefix **'lepido-'** - (uncommon) scaly + trigona (see suffix –trigona) Genus 7

Genus 7 *Lepidotrigona* sp. (Kelulut Kuning) Photo: Azmand Rosland

Lestrimelitta Friese, 1903 (Moure's Bee Catalogue, 2008)= From Old Breton *lestr* (compare Welsh *llestr* ("vessel, receptacle")).**lestr** m (*plural* **listri**) - container, vessel + melitta (see suffix melitta)

Leurotrigona Moure, 1950 (first introduced as a subgenus of *Hypotrigona* Cockerell, 1934) (Moure's Bee Catalogue, 2008) = Ido **le** (*plural*) – **'the'** (*used only when there is no other sign of plurality, for example, with nominalized adjectives*) + uro - Prefix '**-uro'** - (zoology) having a tail + trigona (**see suffix –trigona**)

Liotrigona Moure, 1961 (Eardley & Urban, 2010)= Spanish **lio** (*infinitive* **liar**)- From Latin *ligāre*, present active infinitive of *ligō*. **liar** (*first-person singular present* **lío**, *first-person singular preterite* **lie**, *past participle* **liado**) - to bind, to tie, to wrap, to wrap up + trigona (**see suffix –trigona**)

Genus 8 *Lisotrigona* sp. Photo: Claus Rasmussen

Liotrigonopis Engel 2001 = Baltic amber meliponine genus, Etym.: The new genus-group name is a combination of *Liotrigona* and the suffix *-opsis* (Greek, meaning ''appearing like''). The name references the similarity between the fossil and *Liotrigona,* a genus of African and Malagasy stingless bees. The name is feminine[&&] (Moure J.S. & Camargo J.M.F., 1978)

Lisotrigona Moure, 1961 (Rasmussen, 2008) = From (Portuguese & Spanish) Liso - smooth, plain, flat, straight; From Old Norse *lýsa.* - **lyse** (*imperative* **lys**, *infinitive* **at lyse**, *present tense* **lyser**, *past tense* **lyste**, *past participle* **har lyst**) - shine, give out light + trigona (**see suffix – trigona**) Genus 8

Genus 10 *Melipona* sp. (Urucu boca de Renda) Photo: Lemao Navarro

Lophotrigona Moure, 1961 (Rasmussen, 2008) = From Greek **lophos** – crest - combining form indicating a crested or tufted part: lophophore + trigona (see suffix **–trigona**) Genus 9

Melikerria Moure, (1992). = Latin Meli (honey) + kerria [homage to Dr. Warwik E. Kerr.]* (Moure's Bee Catalogue, 2008)

Warwick Estevam Kerr, the agronomist, geneticist and biologist, is considered the foremost expert in the genetics of bees in the world. (Databank, Warwick Estevam Kerr., 2013) Portrait 29

Meliplebeia Moure, 1961 (Eardley & Urban, 2010) = Latin Meli (honey) + plebeia (see ***Plebeia***)

Melipona Illiger, 1806 (Rasmussen, 2008) = Latin Meli (honey) + pona (Latin "I make") Genus 10

Meliponiden Friese 1903 (Eardley & Urban, 2010) = Melipona (see melipona) + iden (Latin – "identity")

Meliponorytes Tosi 1896 (GBIF, 2013) = Melipona (see melipona) + ory Suffix **-ory** = of or pertaining to, serving for; an adjective suffix appended to various words, often nouns but also verbs, to make an adjective form. + tes (From Latin tensus – "stretched")

Genus 9 *Lophotrigona* sp. Photo by Author.

Meliponula Cockerell (1934) (Eardley & Urban, 2010) = Meli(honey) + ponu (latin "placed by") + la

(Suffix-**lus** m (*feminine* -la, *neuter* -lum); *first/second declension* - Used to form a diminutive of a noun, indicating small size or youth.

Meliwillea Roubik, Lobo & Camargo, 1997 = unusual bee not only for its taxonomic characters and geographical distribution but also because it occurs only within a montane habitat, the cloud forests.; Meli(see **Meli**) + willea - Named after the entomologist Alvaro Wille (see also suffix **-willea**) (Moure's Bee Catalogue, 2008)

Alvaro **Wille** Trejos (San José, Costa Rica, May 17 of 1928 - June 11 of 2006) was an entomologist Costa who devoted his life to studying the insects in Central America bees. Also conducted several fields and was in on the Costa Rican jungles, among which the first study of Corcovado. (Wikipedia, Alvaro Wille, 2012)

Michmelia Moure, (1975). = mich melia (Latin phrase for "much better"); Relates to the subgenus as in *Melipona (Michmelia)* Moure, 1975 - honouring Prof. C. D. Michener (Moure's Bee Catalogue, 2008) + melia (Greek μέλια "honey")

Portrait 2 Charles D. Michener
Photo: Deborah Smith

Micheneria Kerr, et al. (1967). = [homage to Prof. C. D. Michener.]* (Moure's Bee Catalogue, 2008) + ia (see suffix –ia)

The entomologist American Charles Michener was born in Pasadena, California, on 22/9/1918. Much of his career has been devoted to the systematics and natural history of bees. (Wikipedia, Charles Duncan Michener, 2014)

Mourella Moure, (1946) = [homage to Dr. Jesus Santiago Moure]* (Moure's Bee Catalogue, 2008) + ella(see suffix –ella)

Genus 11 *Odontotrigona* sp. Photo: Claus Rasmussen

The first voyages made by Padre Moure abroad occurred in the early 1950s when visiting Argentine institutions. During the stay of Prof. Michener in Brazil, they made a trip to Argentina and Chile. For many years, Father Moure toured throughout Brazil, ministering his famous Numerical Taxonomy courses for students in various research centres and congresses, mainly of Zoology. (Databank, Padre Jesus Santiago Moure, 2013)

Nannotrigona Cockerell, 1922 = From Latin **nānō** -dative singular of *nānus* - From Ancient Greek νᾶνος - **nānus** m (*genitive* **nanī**); *second declension* - dwarf Genus 12

Nogueirapis Moure, 1953 = Named after the entomologist Paulo Nogueira-Neto (Moure's Bee Catalogue, 2008)+ apis(Latin – "bee")

Paulo Nogueira-Neto Professor of Ecology, Institute of Biosciences A.P., University of São Paulo, Brazil Honorary Professor of the Institute of Studies Forward, University of São Paulo (Nogueira-Neto, 1997)

Odontotrigona Moure, 1961 (Rasmussen, 2008) = Origin from Gk odous, odont- 'tooth'.-combining forms relating to a tooth or teeth: odontology. + trigona (see suffix –**trigona**) Genus 11

Oxytrigona Cockerell, 1917 (first introduced as a subgenus of *Trigona* Jurine, 1807) (Moure's Bee Catalogue, 2008) = From the Ancient Greek *ὀξύς* (oxus, "sharp").Prefix '**oxy-**' pungent, acidic (as in the bite of Oxytrigona) + trigona (see suffix **–trigona**)

Papuatrigona Michener & Sakagami, 1990 (Rasmussen, 2008) = Pertaining to Papua New Guinea. + trigona (see suffix –trigona)

Papua (Indonesian: *Provinsi Papua*) is a province of Indonesia. It comprises most of the western half of the island of New Guinea and nearby islands. (Wikipedia, West Papua, 2014)

Paratrigona Schwarz, 1938 (Moure's Bee Catalogue, 2008) = from or pertaining to Pará, a state in the north of Brazil. + trigona (see suffix –trigona)

Parapartamona Schwarz, 1948 (Moure's Bee Catalogue, 2008) = Para (see Latin **para-**) + parta (see **Parta**) + mona (see suffix **–mona**)

Paratrigonoides Camargo & Roubik, 2005 (Moure's Bee Catalogue, 2008) = Para (see Latin **para-**) + trigon (see **Trigona**) + -oides (see suffix **–oides**)

Pariotrigona Moure, 1961 (Rasmussen, 2008) = Latin pario - spawn, produce, procure, acquire + trigon (see **Trigona**)

Partamona Schwarz, 1939 (Moure's Bee Catalogue, 2008) = Parta(see **Parta**) + mona(see suffix –**mona**); having been born of the moon.

Genus 12 *Nannotrigona* sp. (Irai) Photo: Luiz Carlos Da Cunha

Patera Schwarz, (1938) (Moure's Bee Catalogue, 2008) = Latin **patera** f (*genitive* **paterae**); *first declension* - A broad, flat dish or saucer, used especially for libations

Platytrigona Moure, 1961 (Rasmussen, 2008) = English platy - Resembling plates. (heraldry) Of a field semy of plates. + trigona (see **Trigona**)

Plebeia Schwarz, 1938 (Moure's Bee Catalogue, 2008) = Notice from original publication: The name Plebeia was selected a long time ago for this subgenus Dr. F. E. Lutz, but he refrained from publishing it. It signifies undistinguished and denotes the rather commonplace characters that differentiate this subgenus from those subgenera more unusual in their structure. (Schwarz, 1938) **Genus 13**

Genus 13 *Plebia* sp. Fotografía: webbee.org.br via Maria rosa Bernasconi

Plebeina Moure, 1961 (Eardley & Urban, 2010) = From Latin **plebe** – commoner (see also ***Plebeia***) + Latin –ina ("dimunitive form")

Plebeiella Moure, 1961 (Eardley & Urban, 2010) = From Latin **plēbēius** m (feminine plēbēia, neuter plēbēium); first/second declension - plebeian, common, vulgar + ella = Latin Suffix **'-ella'** added to a noun to form a diminutive of that noun.

Proplebeia Michener, (1982). = [An extinct genus *Proplebeia*, recognizes four species, three of them from the amber of Dominican Republic, and one from the amber of Chiapas, Mexico, dating from the Miocene (see Camargo *et al*., 2000)]* (Moure's Bee Catalogue, 2008) Prefix Pro- ("earlier") + plebeia (see plebeia) (first introduced as a subgenus of *Trigona*Jurine, 1807). Fig 1

Ptilotrigona Moure, 1951 (first Introduced to the subgenus of *trigona*. Jurine 1807) (Moure's Bee Catalogue, 2008) = From Greek πτιλο ptilo meaning 'soft feathers' + trigona (see **Trigona**)

Sakagamilla Moure, (1989) = [honouring Soichi F. Sakagami]* (Moure's Bee Catalogue, 2008) + -illa (see suffix **–illa**)

Soichi Francisco **Sakagami** (Born Jan.1927 at Tokyo's Shiba – Died 4/11/1996 in Sapporo) Speciality in Social insects (Senckenberg Biographies, 2013)

Scaptotrigona Moure, (1942) (Introduced first as a subgenus of *Trigona*. Jurine, 1807) (Moure's Bee Catalogue, 2008) = From Latin **scape** - *second declension* - shaft (or similar upright column) + trigona (see **Trigona**) Genus 14

Genus 14 *Scaptotrigona* sp. (Canudo or Mandguari) Photo: Luiz Carlos Da Cunha

Scaura Schwarz, 1938 (first introduced as a subgenus of*Trigona* Jurine, 1807) (Moure's Bee Catalogue, 2008) = Latin - with large ankles

Schwarziana Moure, 1943 = Named after the entomologist Herbert F. Schwarz (Moure's Bee Catalogue, 2008) + iana(see suffix –iana) see also *Schwarzula*

Schwarzula Moure, 1946 = Named after the entomologist Herbert F. Schwarz (Moure's Bee Catalogue, 2008) + ula(see suffix ula) see also *Schwarziana*

Herbert Ferlando Schwarz, an American entomologist, specializing in Hymenoptera, mostly in stingless bees, Meliponini Tribe. He described numerous genera and species. Schwarz was working in the section on insects and spiders the American Museum of Natural History (American Museum of Natural History) (Wikia, Herbert Ferlando Schwarz, 2014)

Sundatrigona Inoue & Sakagami, 1993 (Rasmussen, 2008) = of the Sunda Islands, a group of islands in the western part of the Malay archipelago

Genus 15 *Tetragonisca* sp. (Jatai) Photo: Luiz carlos Da Cunha

Tetragonilla Moure, 1961 (Rasmussen, 2008) = Tetragon ("quadrilateral") + -illa ("small form")

Tetragonisca Moure, 1946 (Moure's Bee Catalogue, 2008) = Tetragon ("quadrilateral") + isca (Latin esca – "bait, lure") Genus 15

Tetragonula Moure, 1961 (Rasmussen, 2008) = Tetragon ("quadrilateral") + ula ("dimunitive form") Genus 16

Genus 16 *Tetragonula* sp. (Bari bari) Photo: Azmand Rosland

Tetrigona Moure, 1961 (Moure's Bee Catalogue, 2008) = Latin te- "thou" + trigona(see **trigona**)

Trichotrigona Camargo & Moure, 1983 = From Ancient Greek *θρίξ* (thríks, "hair"), (stem *trikh-*). **Prefix tricho-** Forming words related to hair + trigona (see **trigona**)

Trigonisca Moure, 1950 = Trigona (see **Trigona**) + isca (From Latin esca) Portuguese **isca** = bait (a substance used in catching fish or other animals) or lure (something that tempts or attracts, especially one with a promise of reward or pleasure)

More Genus Photos

Genus 19 Dactylurina sp. Photo: Peter Kwapong

Genus 18 Meliponula sp. Photo: Peter Kwapong

Genus 17 Friesella sp. Photo: Wilson Gussoni

Other related Genus

Other Genus names that were possibly the wrong combination earlier and some have been placed as Synonyms in Subsequent designation.: [Moure, J. S. (1960). Notes on the types of the Neotropical bees described by Fabricius (Hymenoptera: Apoidea). *Stud. Entomol. (Rio de Janeiro)*]

Alphaneura Gray, G. (1832). = *Class Insecta, arranged by Baron Cuvier, with supplementary additions to each order, and notices of new genera and species by George Gray [in Griffith, E., The animal kingdom arranged in conformity with its organization by Baron Cuvier, member of the Institute of France, etc., with supplementary additions to each order, vol. 15]* Vol. 2. London: Whittaker, Treacher, and Co. 796 pp. [plate 111, figures 1 and 1a] (*junior subjective synonym of Trigona* Jurine, 1807 [**lapsus** for ***Aphaneura*** Gray, 1832 according to Michener, 1994: 375; junior synonym of *Trigona* Jurine according to Michener, 2000: 803]* see **Aphaneura**

Amalthea Rafinesque, C. S. (1815) = In Greek, **Amalthea** or **Amaltheia** (Greek: Ἀμάλθεια) is the most-frequently mentioned foster-mother of Zeus. Her name in Greek ("tender goddess") is an epithet. ; (unnecessary *nom. nov.* for *Trigona* Jurine, 1807 (unjustified replacement, according to Michener, 1997: 6)

Aphaneura Gray, G. (1832). = **Aphane** (From Ancient Greek *ἀφανής* (aphanēs, "indistinct, unseen, obscure"). From English **aphanite** (*plural* **aphanites**) - **Aphanite** - (geology) Certain dark igneous rocks having grain so fine that the individual crystals cannot be seen with the naked.) + ura (Greek ura, "tail").; having an unseen(invisible) tail," Stingless".

Aphaneura rufescens Gray, G. (1832) (junior subjective synonym) under *Trigona pallens* (Fabricius, 1798) http://moure.cria.org.br/catalogue?id=34822

Apis Linnaeus, 1758 = Latin **apis** f (*genitive* **apis**); *third declension* - A bee.

Apis crassipes Fabricius, J. C. (1793) Synonum of *Trigona crassipes* (Fabricius, 1793) http://moure.cria.org.br/catalogue?id=35047 ;

Apis pallida Latreille, P. A. (1804) (junior subjective synonym) and *Apis pallens* Fabricius, J. C. (1798) synonym of *Trigona pallens* (Fabricius, 1798) http://moure.cria.org.br/catalogue?id=34822 ;

Apis ruficrus Latreille, P. A. (1804) (junior subjective synonym) and *Apis spinipes* Fabricius, J. C. (1793) synonym of *Trigona spinipes* (Fabricius, 1793) http://moure.cria.org.br/catalogue?id=34822 ;

Apis flavicornis Fabricius, J. C. (1798) Synonym of *Frieseomelitta flavicornis* (Fabricius, 1798) http://moure.cria.org.br/catalogue?id=34822 ;

Apis compressipes Fabricius, J. C. (1804) Synonym of Melipona (Melikerria) compressipes (Fabricius, 1804) http://moure.cria.org.br/catalogue?id=34750 ;

Apis atrata Fabricius, J. C. (1798) (junior subjective synonym) and *Apis favosa* Fabricius, J. C. (1798) Synonym of *Melipona (Melipona) favosa* (Fabricius, 1798) http://moure.cria.org.br/catalogue?id=34750

Apis sylvestris Barrère, P. (1741) (*Nomen nudum* ? [see Schwarz, 1948: 243, 254]) and *Apis amalthea* Olivier, A. G. (1789) Synonym of Trigona amalthea (Olivier, 1789) http://moure.cria.org.br/catalogue?id=34854

Anthidium Fabricius, 1804 = From Ancient Greek *ἄνθος* (anthos) – Latin anthus m (*genitive* **anthī**); *second declension* - a small bird, possibly the yellow wagtail. + Latin dius m (*feminine* dia, *neuter* dium); ***first/second declension*** - Alternative form of **dīvus** m (*feminine* **dīva**, *neuter* **dīvum**); *first/second declension* - of or belonging to a deity; divine.

Anthidium pictifrons Packard, A. S., Jr. (1869).(junior subjective synonym) to *Melipona (Melikerria) grandis* Guérin, 1844 http://moure.cria.org.br/catalogue?id=34751

pictifrons - **{picti** = Latin **pictī** - nominative masculine plural of *pictus* - **pictus** m (*feminine* picta, *neuter* pictum); *first/second declension* - decorated, embellished, having been decorated}+**{frons** = the upper anterior portion of the head of an insect, above or behind the clypeus; forehead.}

Bombus Latreille, 1802 = From Ancient Greek *βόμβος* (bombos) - Latin **bombus** m (*genitive* **bombī**); *second declension* - A buzz or humming sound.

Bombus parvulus Fabricius, J. C. (1804). (junior subjective synonym) to *Melipona (Melipona) favosa (*Fabricius, 1798) http://moure.cria.org.br/catalogue?id=34629

Latin **parvulus** m (*feminine* parvula, *neuter* parvulum); *first/second declension* - little, petty, slight.

Centris Fabricius, 1804 = From Latin **centrum** n (genitive centrī); second declension - centre/center + -s (Suffix **'-s'** (not productive) used in the formation of certain English adverbs.). *Centris cilipes* Fabricius, 1804 by original designation, now a synonym of *Trigona cilipes* (Fabricius, 1804) http://moure.cria.org.br/catalogue?id=34843 ;

Centris pediculana Fabricius, J. C. (1804) synonym of *Trigonisca pediculana* (Fabricius, 1804) http://moure.cria.org.br/catalogue?id=35047

Dialictus nigrocyaneus (Ashmead, 1900) (APIDAE ›› Halictinae ›› Halictini ›› Halictina)

Trigona nigrocyanea Ashmead, W. H. (1900). **Syn.** Report upon the aculeate Hymenoptera of the Islands of St. Vincent and Grenada, with additions to the parasitic Hymenoptera and a list of the described Hymenoptera of the West Indies. *Trans. Entomol. Soc. Lond.* **2:** 207-367 [208] **nigrocyanea** Etym.: Latin nigro ("the colour black") + Latin **cȳaneus** - deep or dark blue (colour)

Pisania Cunha, R. A. (1973).= Latin pisa- nominative plural of *pīsum* -From Ancient Greek *πίσον* (pison).**pīsum** n (*genitive* **pīsī**); *second declension* - pea (*plant*) + nia (Latin nia, "inhibit"); "pea-sized inhabitants." ;

Pisania Cunha, R. A. (1973). Taxonomia numérica de alguns Meliponinae (Hymenoptera- Apidae). *Ciência Biológica (Portugal)* **1:** 25-42 [40 (reprint, page 16)] (first introduced as a subgenus of *Lestrimelitta* Friese, 1903, monobasic, including only *Trigona cilipes* (Fabricius, 1804))

Chapter 4

Patronymics

In English, Patronym is from Ancient Greek πατήρ (*patēr*, “father”) + ὄνομα (*onoma*, “name”). The word "patron" derives from the Latin *patronus*, "patron".

In biological taxonomy, a patronym is the second part of a binomial derived from a Latinized surname. These often honour associates of the biologist who named the organism rather than the authors themselves.

This section lists meliponine nomenclature that pays for naturalists, specimen collectors, zoologists, entomologists and explorers.

Description Indicators

Type Locality *(Noun) abbr.Typ.Loc.*

The location, usually where the first specimen was discovered, is considered the most representative of a formation.

Etymology *(n.)abbr Etym.*

The study of the history of words or element of a word and their origins

Patronymics of Meliponine Species

Suffixes and Latin terminates used in this section:

-ae = used as singular forms within the genitive case belonging to Latin first-declension feminine nouns. (Most Matronyms will have the Latin terminate -ae)

-ana = From Latin **ana** - old (of female persons and things), aged; **From Latin suffix -ānus** m (*feminine* **-āna**, *neuter* **-ānum**); *first/second declension* - Of or pertaining to; usually indicates a relationship of position, possession, or origin. (Some Matronyms may have this suffix)

-iana = Latin Suffix –**iana** = (New Latin) used to form taxonomic names, often from non-Latin stems

-rum = From Latin **rum** meaning “**of**”

Meliponine Species Patronyms & Matronyms:

abrassarti, *Trigona* Cockerell: in Darchen 1971b – and *Apotrigona abrassarti* (Cockerell) syn.: Named after naturalist Dr. Abrassart (Eardley & Urban, 2010)

Abrassart (Maurice-Alfred-Marie) Doctor of medicine and director of societies (Leuven, 16/08/1872 - Brussels, 02/03/1951) (Jadot, 1968)

ailyae, *Partamona* Camargo, 1980 [*Etym.:* Matronym for Dair Aily F. Camargo.]* (Moure's Bee Catalogue, 2008) + suffix -ae

Dair **Aily** Franco de Camargo is Associate Professor, Department of Education - Institute of Biosciences - UNESP - Campus DERIO course. (UNESP, 2014)

alinderi, *Melipona* Alfken, 1929 Syn. n. and [*Trigona (Meliponula)* ***alinderi*** var. *nigrita* (Alfken): in Cockerell 1934a]^ (Eardley, 2004) – Named after entomologist Sven Alinder (1900-1928) (Senckenberg Biographies, 2013)

alfkeni, *Trigona* Friese, H. (1900) [*Etym.:* Patronym for Prof. Dr. Johann Dietrich Alfken.]* (Moure's Bee Catalogue, 2008)

Johann Dietrich **Alfken** (Jun. 11 1862, Frankfurt - Feb. 14 1945, Rüthersdorf, Berlin)was a German entomologist who specialized in Hymenoptera , especially Apoidea. (Senckenberg Biographies, 2013)

aliceae, *Homotrigona* (Cockerell, 1929) [Typ. Loc.: "Thailand: On the trail in the jungle between Pahtoop mountain and Nan, Jan. 9, 1928" (Rasmussen, 2008) Sp. 1

Location description by **Alice** Mackie (accompanied Cockerell on his travels) regarding Trigona aliceae)"[1] (Rasmussen, Personal communication, 2013)

Sp. 1 *Homotrigona aliceae* Photo: Hisham Ismail

alvarengai, *Plebeia* Moure, 1994 [*Typ. Loc.:* Jacareacanga, Pará, Brazil, X.1959, M. Alvarenga leg. *Etym.:* patronym for Moacyr Alvarenga.]* (Moure's Bee Catalogue, 2008)

anduzei, *Paratrigona* (Schwarz, 1943) [*Typ. Loc.:* Tabay, E. Mérida, by P. J. Anduze.]* (Moure's Bee Catalogue, 2008)

araujoi, *Hypotrigona* (Michener), - [Named after the naturalists Professor **Virgílio** de Portugal Araújo;]^ (Eardley, 2004)

asilvai, *Melipona (Eomelipona)* Moure, 1971 [*Etym*.: patronym for Dr. Aristóteles Araújo de Godofredo Silva.]* (Moure's Bee Catalogue, 2008)

atahualpa, *Trigona* Pedro & Camargo, 2003 [Nomen nudum]* (Moure's Bee Catalogue, 2008) - Named after **Atahualpa** (Quechua) (20/3/1497–29/8/1533) was the last Sapa Inca (sovereign emperor) of the Tawantinsuyu (the Inca Empire) before the Spanish conquest. (Wikipedia, Atahualpa, 2014) Portrait 11

Sp. 2 *Melipona beechei* Photo: via Maria Rosa Bernasconi

baeri, *Melipona (Melipona) baeri* Vachal, 1904 - Named after the naturalists/collector Gustave Adolphe **Baer**; [Ref: "Voyage de M. G. A. Baer au Tucuman (Argentine). Hymenoptera Mellifera (Familia única: Apidae). *Rev. Entomol. (Caen)* **23**"] * (Moure's Bee Catalogue, 2008)

bakeri, *Heterotrigona* (Cockerell, 1919) # (Rasmussen, 2008)- Named after the entomologist Edward Charles Stuart Baker, the collector of the species " Malaysia, Penang Island (Baker 9068)"[1] (Rasmussen, Personal communication, 2013) Portrait 11

batesi, *Partamona* Pedro & Camargo, 2003 [*Etym*.: Patronym for Henry Walter Bates.]* (Moure's Bee Catalogue, 2008);

Henry Walter **Bates** FRS FLS FGS (Leicester, 8/2/1825 – London, 16/2/1892) was an English naturalist and explorer who gave the first scientific account of mimicry in animals. He was most famous for his expedition to the rainforests of the Amazon. (Wikipedia, Henry Walter Bates, 2014) Portrait 11

beccarii, *Trigona* Gribodo, 1879 - Named after the entomologist Giovanni **Beccari** ; Relates to [*Meliponula (Meliplebeia)* ***beccarii*** Friese]^ (Eardley, 2004)

beebei, *Tetragona* (Schwarz, 1938) - Named after the naturalists C. W. **Beebe**; Also relates to [*Trigona* Schwarz, 1938]* (Moure's Bee Catalogue, 2008)

Charles William **Beebe** (29/7/1877 – 4/6/1962) was an American naturalist, ornithologist, marine biologist, entomologist, explorer, and author. (Wikipedia, William Beebe, 2014) Portrait 11

beecheii, *Melipona*Bennett, 1831 [Etym.: patronym for Captain Frederick William Beechey.]* (Moure's Bee Catalogue, 2008) Sp. 2

Frederick William **Beechey** (17/2/1796 – 29/11/1856) was an English naval officer and geographer. *Publication:* F.W. Beechey, *Narrative of a Voyage to the Pacific and Behring's Strait" (1831); The Zoology of Captain Beechey's voyage to the Pacific and Behring's Strait.* (1839). (Wikipedia, Frederick William Beechey, 2014) Portrait 11

binghami, *Tetrigona* (Schwarz, 1937): - Named after the entomologist Charles Thomas **Bingham**; Relates to the collector of [*Trigona apicalis* variety *binghami* Schwarz 1937 Typ. Loc.: Myanmar "Tenasserim: Dawnat Range, Jan., 1891 (Col. C.T. Bingham)"][#] (Rasmussen, 2008) Sp. 3

Sp. 3 Tetrigona binghami Photo: Hanif Redland

Charles Thomas **Bingham** (16/4/1848, Ireland – 18/10/1908 West Kensington, London) was an Irish military officer and entomologist. (Wikipedia, Charles Thomas Bingham , 2013) Portrait 11

biroi, *Trigona* Friese 1898 - Named after the naturalists Lajos **Biro**.; Relating to the collector of [*Tetragonula* ***biroi*** (Friese, 1898)][#] (Rasmussen, 2008) Sp. 4

Sp. 4 *Tetragonula biroi* Photo: via Texter Taclobao

Lajos **Bíró** (29/8/1856 – 2/9/1931) was a zoologist from Hungary. (Senckenberg Biographies, 2013)

bocandei, Trigona (Spinola): - Named after the naturalists E. Bertrand **Bocande**; Relates to [*Meliponula (Meliponula)* ***bocandei*** (Spinola), This species is easy to identify as it is the only large stingless bee with orange vestiture, a largely black face, yellowish-orange scutellum and spoon-shaped hind tibia.]^ (Eardley, 2004)

Emmanuel Bertrand-**Bocandé** (1812–1881) was an explorer, businessman, and French colonial administrator. (Wikipedia, Emmanuel Bertrand-Bocandé , 2013)

bottegoi, *Trigona* Magretti, 1895: - Named after the naturalists/collector Vittorio **Bottego** ; Relates to [*Liotrigona* ***bottegoi*** *(*Magretti), The type material of L. bottegoi was collected in Ethiopia]^ (Eardley, 2004)

Vittorio Bottego (29/7/1860 – 17/3/1897) was an Italian army officer and one of the first Western explorers of Jubaland in the Horn of Africa. (Wikipedia, Vittorio Bottego , 2014)

bouyssoui, *Trigona* (Vachal) - Named after the naturalists/collector Joseph **Bouyssou** ; Also relates to [*Melipona (Trigona)* ***bouyssoui*** Vachal, 1903]^ (Eardley, 2004)

bradleyi, *Melipona (Eomelipona)* Schwarz, 1932, [Etym.: in honour of Professor James Chester **Bradley**.]* (Moure's Bee Catalogue, 2008)

branneri, *Trigona* Cockerell, 1912 - Named after the naturalist John Casper **Branner**; Relates to [*Trigona* ***branneri*** Cockerell, 1912]*. (Moure's Bee Catalogue, 2008) The species was described in a paper on the Standford expedition (J C Branner was the leader of the expedition) 1911 to Brazil (maybe not the collector, but formed part of the group that went to Brazil and collected the bee).[1] (Rasmussen, Personal communication, 2013)

John Casper **Branner** (4/7/1850 –1/3/1922) was an American geologist and academic who discovered bauxite in Arkansas in 1887. He was Chair of the Departments of Botany and Geology at Indiana University and later at Stanford University. He served as President of the Geological Society of America in 1904. (Wikipedia, John Casper Branner , 2014)

braueri, *Trigona* Friese, 1900 [*Etym.:* patronym for Prof. Fr. **Brauer.**]* (Moure's Bee Catalogue, 2008)

Friedrich Moritz Brauer (12/5/1832, Vienna – 19/12/1904). An Austrian entomologist appointed Professor of zoology in the University of Vienna; He was elected in an Honorary Fellow of the Entomological Society of London in 1900 (Wikipedia, Friedrich Moritz Brauer , 2013)

braunsi, *Hypotrigona* Kohl - Named after the entomologist Sigismund **Brauns**; Relates to [*Trigona* ***braunsii*** Kohl, 1894. *Melipona (Trigona)* ***braunsi*** [!] (Kohl): Vachal 1903. *Hypotrigona* ***braunsi*** [!] (Kohl): Friese 1909a. *Melipona* ***braunsi*** [!] (Kohl): Loveridge 1923. *Trigona (Hypotrigona)* ***braunsi*** [!] Kohl: Araujo & Kerr 1959. *Hypotrigona* ***braunsi*** [!] (Kohl): Moure 1961: the name **braunsi** (spelt with one 'i') appears more often in the literature because of the spelling discrepancy in **braunsii.]**^ (Eardley, 2004)

braunsii, *Trigona* Kohl, 1894. See also **braunsi**

Sigismund Brauns (2/4/1839 Oesselse, Hanover – 27/3/1913 Schwerin) A school teacher in Schwerin, speciality in Hymenoptera and Coleoptera. (Senckenberg Biographies, 2013)

browni, *Dolichotrigona* Camargo & Pedro, 2005. [Etym.: patronym for John Christopher **Brown**.]* (Moure's Bee Catalogue, 2008)

buchwaldi, *Tetragonisca* (Friese, 1925) [*Typ. Loc.:* Guayaquil, Ecuador, I.1909, by Frederico von **Buchwald** leg..]* (Moure's Bee Catalogue, 2008)

F. v. Buchwald, Collection: Insects of Quivedo, Ecuador in 1903 and 1914 WFH Rosenberg/London (Senckenberg Biographies, 2013)

buschirii, *Axestotrigona* (Friese) - Patronym to Buschiri an entomologist and insect collector in Cameroon 1912 (Eardley & Urban, 2010)

butteli, *Nogueirapis* (Friese, 1900) [*Etym.:* patronym for H. v. **Buttel**-Reepen.]* (Moure's Bee Catalogue, 2008)

Hugo von Buttel-Reepen (02/12/1860 – 1933) was a Zoologist and Director of Oldenburg's natural history Museum. (Senckenberg Biographies, 2013)

buyssoni, *Trigona* Friese, 1902 - Named after the entomologist Robert Du **Buysson**.; Relates also to [*Trigonisca* ***buyssoni*** (Friese, 1902)]* (Moure's Bee Catalogue, 2008)

Robert Du **Buysson** (1861-1946) speciality in Hymenoptera and Coleoptera (Senckenberg Biographies, 2013)

cacciae, *Melipona* Nurse 1907 syn. - Relates to [*Lisotrigona* ***cacciae*** (Nurse, 1907)][#] (Rasmussen, 2008) Charles George Nurse (1862–1933) based his description on specimens collected by A.M.F. Caccia, of the Indian Forest Department, India. (Rasmussen, 2008) + ae(see suffix **-ae**) Sp. 5

Sp. 5 *Lisotrigona cacciae* Photo: Claus Rasmussen

camargoi, *Camargoia* Moure, 1989; [Etym.: patronym for Prof. João Maria Franco de Camargo.]* (Moure's Bee Catalogue, 2008)

João Maria Franco Camargo: A Brazilian ecologist and bees taxonomist working on Tribe Meliponini. (Wikia, João M. F. de Camargo, 2007) Portrait 20

carpenteri, *Lisotrigona* Engel 2000 - Relates to the collector of *Lisotrigona* ***carpenteri*** Engel, 2000 [Typ. Loc.: Pozo Vietnam: Nghê An Province: Khe Bo, by James Michael Carpenter from the AMNH][#] (Rasmussen, 2008)

carrikeri, *Melipona* Cockerell, 1919 - Refers to the collector of *Melipona (Eomelipona)* ***carrikeri*** Cockerell, 1919 [Typ. Loc.: Pozo Azul, Costa Rica, by M. A. **Carriker.]*** (Moure's Bee Catalogue, 2008)

Melbourne Armstrong **Carriker** (1879-1965) was an American ornithologist and entomologist (Wikipedia, Melbourne Armstrong Carriker , 2013)

cockerelli, *Trigona* Rayment 1930 - Named after the entomologist T. D. A. **Cockerell**; Relates to [*Austroplebeia* ***cockerelli*** (Rayment, 1930) and *Trigona* ***cockerelli*** *ornata* subsp. nov. Rayment 1932f][#] (Rasmussen, 2008)

Theodore Dru Alison **Cockerell** (1866–1948) was an American zoologist, born in Norwood, England, and brother of Sydney Cockerell. He was educated at the Middlesex and then studied botany in Colorado in 1887-90. Subsequently, he became a taxonomist and published numerous papers on the Hymenoptera, Hemiptera, and Mollusca and publications on palaeontology and evolution. (Wikipedia, Theodore Dru Alison Cockerell , 2013) Portrait 29

conradti, Trigona Friese, 1900^ = Named after entomologist Leopold Conradt (Eardley, 2004)

cramptoni, *Melipona* Cockerell, 1920 - Relates to the collector of *Melipona (Michmelia)* ***cramptoni*** Cockerell, (1920). [*Typ. Loc.:* Paiapalu Creek, northern Brazil, by H. E. Crampton & F. E. Lutz]* (Moure's Bee Catalogue, 2008)

Henry Edward **Crampton** (5/1/1875 - 26/2/1956) was an American palaeontologist and evolutionary biologist. (Wikipedia, Henry Edward Crampton , 2013)

cressoni, *Melipona* Dalla Torre (1896). (unnecessary nom. nov. for *Trigona nigra* Cresson, 1878)* - Named after the naturalists named E. T. Cresson Sr. (Moure's Bee Catalogue, 2008)

Ezra Townsend Sen. Cresson (Born18/6/1838 in Byberry, PA – Died 19/4/926 in Swarthmore, PA) Hymenoptera. (Senckenberg Biographies, 2013)

curriei, *Lestrimelitta (Cleptotrigona)* (Cockerell): Moure 1961 and syn.n *Trigona* ***curriei*** Cockerell, 1917^: = Named after entomologist R.P. Currie. (Eardley, 2004)

Rolla Patterson Currie (Born 25/3/1875 in Preemption, IL – Died 1960) Speciality in Lepidoptera. (Senckenberg Biographies, 2013)

dallatorreana, *Trigona* Friese, 1900 [*Etym.:* patronym for Prof. K. W. Dalla Torre.] * (Moure's Bee Catalogue, 2008) + suffix –ana. Sp. 6

Karl Wilhelm von **Dalla Torre** (14/7/1850 – 6/4/1928) was an Austrian taxonomist, entomologist and botanist. Dalla Torre was born in Kitzbühel, Tyrol. He studied natural sciences at the University of Innsbruck. He then worked in the University as an entomologist and in 1895 became Professor of Zoology at the University of Innsbruck (Wikipedia, Karl Wilhelm von Dalla Torre , 2014)

danuncia, *Lestrimelitta* Oliveira & Marchi, 2005 Etym.: homage to Dra. Danuncia Urban.]* (Moure's Bee Catalogue, 2008)

Sp. 6 *Trigona dallatorreana* Photo: Claus Rasmussen

Dr. Danúncia Urban, a prominent researcher of wild bees in the Department of Zoology., Universidade Federal do Parana. (UFPR, 2012)

davenporti, *Tetragonula* (Franck, 2004) also relating to *Trigona (Heterotrigona) davenporti* **Franck *et al.*** 2004 [Typ. Loc.: Australia, Qld, Mudgeeraba, Austinville Road, V. 1990, P. Davenport leg."][#] (Rasmussen, 2008)

This species was dedicated to Peter **Davenport,** who collected the four colonies examined.[1] (Rasmussen, Personal communication, 2013) Portrait 20

denoiti. *Melipona (Trigona)* Vachal, 1903:syn. n. of *Plebeina* ***denoiti*** (Vachal): in Moure 1961 = Named after entomologist R. P Denoit. (Eardley & Urban, 2010)

[likely Reverend Pastor Camille Denoit] (1862–1891): Uganda (Rasmussen, Biography Joseph Vachal, 2012)

dobzhanskyi, *Trigonisca* (Moure, 1950) [*Typ. Loc.:* Belém, Pará, Brazil, 8-10. V.1949, T. **Dobzhansky** leg.. *Etym.:* patronym for Prof. Theodosius **Dobzhansky**.]* (Moure's Bee Catalogue, 2008)

Sp. 7 *Plebeia drotyana* Photo: Luiz Carlos Da Cunha

Theodosius Grygorovych **Dobzhansky** (24/1/1900 – 18/12/1975) was a prominent geneticist and evolutionary biologist and a central figure in evolutionary biology. (Wikipedia, Theodosius Dobzhansky, 2014) Portrait 35

doederleini, *Frieseomelitta* (Friese, 1900) [Etym.: patronym for Prof. L. **Doederlein**.]* (Moure's Bee Catalogue, 2008)

Ludwig Heinrich Philipp Döderlein, was a zoologist and paleontologist (1855–1936) (Senckenberg Biographies, 2013)

dominiquella, *Trigona* Dominique, J. (1898) (Nomen nudum) .. Named by author Entomologist Jules Dominique in his name (Moure's Bee Catalogue, 2008) + ella (se suffix –ella)

Jules Dominique was an entomologist (Born 9/1838 in Nevers – Died 7/12/1902 in Nantes) Speciality in Coleoptera. (Senckenberg Biographies, 2013)

drescheri, *Tetragonula* (Schwarz, 1939) [Typ. Loc.: Indonesia -M. Java.-South Banjoemas, Koebang kangkoeng, July 1935 by Friedrich Carl Drescher][#] (Rasmussen, 2008)

Drescher, Friedrich Carl (1875-1957) Plant collector in Malenesian Region. (Jstor, 2013)

droryana, *Trigona*Friese, 1900 - Named after the entomologist E. Drory + ana(see suffix – ana); In relation with [*Plebeia* ***droryana*** (Friese, 1900)]* (Moure's Bee Catalogue, 2008) Sp. 7

One of the key number of colonies involved was Drory in France, 1872-1873, in the Bordeaux region. Bees were sent from Bahia by the French naturalist Louis Jacques Brunet, who lives in the Northeast. Unfortunately, the bees did not survive for long the rigours of European winter but allowed the realization of important observations by Edouard Drory, Maurice Girard and others. (Nogueira-Neto, 1997)

duckei, *Trigonisca* (Friese, 1900) [*Typ. Loc.:* Belém, Pará, Brazil by A. **Ducke** leg.*.Etym.:* patronym for Adolf **Ducke**.]* (Moure's Bee Catalogue, 2008)

Adolpho Ducke (19/10/1876 in Trieste, Italy – 5/1/1959 in Fortaleza, Brazil), also referred to as **Adolfo Ducke** and occasionally misspelt **"Duque"**, was a notable entomologist, botanist and ethnographer of Amazonia. (Wikipedia, Adolpho Ducke, 2014)

dudgeoni, *Axestotrigona* Moure, 1961- Named after the entomologist Gerald C. Dudgeon; In relation with [*Axestotrigona dudgeoni* Moure, 1961]^ (Eardley, 2004)

Gerrald C. Dudgeon died in 1930, speciality in Lepidoptera and Hymenoptera. Collection: Lepidoptera: Heterocera and Hymenoptera, Aculeata 1904, Ind. Mus. Calcutta; African Rhopalocera of Brit. Mus. (NH) London. (Senckenberg Biographies, 2013)

duponti, *Trigona* Spinola, M. (1840) (Nomen nudum) – Named after the entomologist R.H. Dupont. (Moure's Bee Catalogue, 2008)

Richard Henry Puech Dupont (Born 11/01/1798 in Bayeux, Calvados – Died 07/02/1873 in Bellevue, Seine-et-Oise) Speciality in Coleoptera. (Senckenberg Biographies, 2013)

dutrae, *Nannotrigona* (Friese, 1901) [*Etym.:* Homage to Dr. **Dutra**.]* (Moure's Bee Catalogue, 2008) + suffix **-ae**

emersoni, *Scaptotrigona* (Schwarz, 1938) [*Typ. Loc.:* Kartabo, British Guiana, 13.III.1924, by A. **Emerson**.]* (Moure's Bee Catalogue, 2008)

Alfred Edwards Emerson (31/12/1896 – 3/10/1976) was an American biologist, Professor of Zoology at the University of Chicago, a noted entomologist and leading authority on termites. (Wikipedia, Alfred E. Emerson, 2013)

ehrhardti, *Trigona* Friese, 1931 - Named after the entomologist W. Ehrhardt also relates to [*Lestrimelitta* ***ehrhardti*** (Friese, 1931)]* (Moure's Bee Catalogue, 2008)

euclydiana, *Celetrigona* Camargo & Pedro, 2009 [Etym.: Patronym for **Euclydes** Rodrigues Pimenta da Cunha, author of Os Sertões, who participated, in 1905, in the Comissão Mista Brasileiro-Peruana de reconhecimento do Alto Purus, Acre, where *C. euclydiana* is found.]* (Moure's Bee Catalogue, 2008) + ana (see suffix –ana)

ferreirai, *Partamona* Pedro & Camargo, 2003 [*Etym.:* patronym for Alexandre Rodrigues **Ferreira.]*** (Moure's Bee Catalogue, 2008)

Alexandre Rodrigues **Ferreira** (27/4/1756 – 23/4/1815) was a naturalist born in the Portuguese colony of Brazil. He undertook an extensive journey that crossed the interior of the Amazon Basin to Mato Grosso between 1783 and 1792. (Wikipedia, Alexandre Rodrigues Ferreira , 2014)

fiebrigi, *Tetragonisca* (Schwarz, 1938) [*Typ. Loc.:* San Bernardino, Paraguay, collected by K. **Fiebrig**.]* (Moure's Bee Catalogue, 2008)

Fiebrig, Karl (Carl) August Gustav 1869–1951 (Senckenberg Biographies, 2013)

fraissei, *Trigonisca* (Friese, 1901) [*Etym.:* patronym for Prof. Paul Hermann **Fraisse**.]* (Moure's Bee Catalogue, 2008)

Fraisse, *Paul* Hermann (Born 4/12/1851 in Memel (former East Prussia) – Died 3/11/1909 in Jena) was a Zoologist and Explorer. (Senckenberg Biographies, 2013)

franki, *Plebeia* (Friese, 1900) [*Etym.:* patronym for A. **Frank**.]* (Moure's Bee Catalogue, 2008)

francoi, *Frieseomelitta* (Moure, 1946) [*TYP. LOC.:* Riachuelo, Sergipe, Brazil, A. **Franco** Filho leg. Etym.: patronym for Dr. A. **Franco** Filho.]* (Moure's Bee Catalogue, 2008)

freiremaiai, Frieseomelitta [(Moure, 1963) [Etym.: patronym for Drs. Newton Freire-Maia and Ademar Freire-Maia.]* (Moure's Bee Catalogue, 2008)

friesei, *Trigona* Ihering, 1903 -Named after the entomologist Heinrich **Friese;** Relates to [*Trigona* ***friesei*** Ihering, H. (1903)]* (Moure's Bee Catalogue, 2008)

Heinrich Friedrich August Karl Ludwig **Friese** (Born 4/5/1860 in Schwerin - Died 8/9/1948 in Schwerin) was a German biologist and entomologist, specialist of bees. (Senckenberg Biographies, 2013) Portrait 11

geissleri, *Trigona* **Cockerell** 1918: - Named after the entomologist K. Geissler.also in relation to [*Tetragonula* ***geissleri*** (Cockerell, 1918)][#] (Rasmussen, 2008)

ghilianii, *Duckeola* (Spinola, 1853) [Account of coming from unpublished Hymenoptera entomological trip Mr. Ghiliani in Para in 1846.]*

Vittore Ghiliani (1812-1878) was an Italian entomologist from Torino speciality in Coleoptera, Lepidoptera. (Senckenberg Biographies, 2013)

goeldiana, *Plebeia* (Friese, 1900) [*Etym.:* patronym for Dr. E. Göldi.]* (Moure's Bee Catalogue, 2008) with the Latin suffix *-ana* ('a connection')

Émil August **Goeldi** (var. **Göldi**, var. Emílio Augusto **Goeldi**) (August 28, 1859 – July 5, 1917 in Bern), was a Swiss-Brazilian naturalist and zoologist. He was the father of Oswaldo Goeldi, a noted Brazilian engraver and illustrator. (Wikipedia, Émil Goeldi, 2014) Portrait 29

graeffei, *Trigonisca* (Friese, 1900) [*Etym.:* patronym for Dr Eduard Heinrich Graeffe.]* (Moure's Bee Catalogue, 2008)

Eduard Heinrich **Graeffe** or **Gräffe** (27/12/1833, Zurich – 23/4/1916 Ljubljana) was a Swiss zoologist and naturalist. As an entomologist, he specialised in Hymenoptera, Diptera and Hemiptera. (Wikipedia, Eduard Heinrich Graeffe, 2013)

gressitti, Trigona (Tetragonula) Sakagami 1978 - Named after Judson Linsley Gressitt (1914-1982) and in relation with [*Tetragonula gressitti* (Sakagami, 1978)][#] (Rasmussen, 2008)

Judson Linsley Gressitt was born in Tokyo in 1914 to an American family. He grew up in Tokyo and was educated at an American school. On finishing school, he travelled alone at age 18 to Taiwan and collected insects in much of the island, including the highest mountains. His first degrees were in entomology at the University of California. (Bishop Museum , 1983)

gribodoi, *Trigona* Magretti, 1884: - Named after the entomologist Giovanni Gribodo.; Refers to [Hypotrigona **gribodoi** (Magretti), gribodoi is the preferred choice because of the spelling discrepancy in braunsii.]^ (Eardley, 2004)

Giovanni Gribodo was born in 1846 and died in 1924 in Turin. Graduated in 1866 from the Application School for Engineers of Turin, Gribodo was an assistant of Practical Geometry and Applied Descriptive Geometry. A fan of entomology and architecture. (Virtual Museum, 1998-99)

griswaldorum, *Meliponula (Meliplebeia) griswoldorum sp.n.*, [Etym.: This species is named for Terry and Rhonda Griswold, who collected the type material near Chipata, Zambia.]^ (Eardley, 2004) + rum(From Latin **rum** meaning "**of**")

goettei, *Tetragona* (Friese, 1900) [*Etym.:* patronym for Prof. Alexander W. von Goette.]* (Moure's Bee Catalogue, 2008)

Alexander W. von **Götte** (31/12/1840 – 5/2/1922) was a Russian born zoologist, born in St. Petersburg. (Wikipedia, Alexander W. von Götte , 2013)

guigliae, *Paratrigona* Moure, 1960 - Named after the naturalists Delfa Guigli also relates to [*Paratrigona (Paratrigona) ornaticeps* ***guigliae*** Moure, 1960]* (Moure's Bee Catalogue, 2008) + Latin terminate **-ae** (see suffix **–ae)**

handlirschii, *Tetragona* (Friese, 1900) [Etym.: patronym for A. Handlirsch.]* (Moure's Bee Catalogue, 2008)

Anton Handlirsch (1865 in Wien -1835 in Wien) was a Professor and Director of Natural History Museum in Wien. Specialist in Hymenoptera, Hemiptera (Senckenberg Biographies, 2013)

haeckeli, *Trigona* Friese, 1900 - Named after the zoologist E. Haeckel in relation with [***Paratrigona haeckeli*** (Friese, 1900)]* (Moure's Bee Catalogue, 2008)

Ernst Heinrich Philipp August Haeckel (Born 16/2/1834 in Potsdam – Died 11/8/1919 in Jena) Physicians, Zoologist and Museum Director in Jena. (Senckenberg Biographies, 2013)

helleri, *Partamona* (Friese, 1900) [*Etym.:* patronym for Prof. C. **Heller**.]* (Moure's Bee Catalogue, 2008)

Prof. Dr. Camille **Heller** (Born 1823 in Sobochleben in Teplice – Died 1917) Speciality Hymenoptera. (Senckenberg Biographies, 2013)

hellwegeri, *Trigona* Friese - Named after the naturalist Michele **Hellweger** in relation with [*Scaptotrigona* ***hellwegeri*** (Friese, 1900)]* (Moure's Bee Catalogue, 2008)

Michele **Hellweger** (Born 1/2/1865 in Innsbruck – Died 28/3/1930 in Bressanone, Bolzano) a Pastor and naturalist, speciality in Lepidotera. (Senckenberg Biographies, 2013)

heideri, *Trigona* Friese, H. (1900) [*Etym.:* patronym for Prof. C. Heider.]* (Moure's Bee Catalogue, 2008)

hildebrandti, *Trigona* Friese, 1900: - Named after the naturalists **Hildebrandt** and in relation to [*Plebeina* ***hildebrandti*** (Friese) comb.n.,]^ (Eardley, 2004)

Johannes **Hildebrandt** (1847-1881) was a German explorer and collector. (Wikipedia, Johannes Hildebrandt, 2013)

hirashimai, *Tetragonula* **(**Sakagami, 1978) **-** Named after the naturalists Yoshihiro Hirashima and in relation with [*Trigona (Tetragonula)* ***hirashimai*** Sakagami 1978][#] (Rasmussen, 2008)

hobbyi, *Trigona* Schwarz 1937 - Named after the entomologist named Bertram Maurice **Hobby** in relation with *[Platytrigona* ***hobbyi*** (Schwarz, 1937)][#] (Rasmussen, 2008)

Dr. Bertram Maurice Hobby (Born 23/10/1905 in Southampton – Died 19/7/1983) Speciality in Diptera. (Senckenberg Biographies, 2013)

hockingsi, *Tetragonula* (Cockerell, 1929) - Named after the collector of *Trigona carbonaria* ***hockingsi*** n. subsp. Cockerell 1929 **[**Typ. Loc.: AUSTRALIA **the** Cape York Peninsula by Harold **Hockings]**[#] (Rasmussen, 2008) **Sp. 8**

Harold Joseph **Hockings** (Born1856 - D.) Birthplace, Thursday Island. He was a keen naturalist and geologist and made a special study of native bees, specimens he forwarded to many scientific bodies in England and America. (Geni, 2014)

Sp. 8 *Tetragonula hockingsi* Photo: Robert Luttrel

huberi, *Trigona* Friese, 1901 [Etym.: patronym for Dr. J. **Huber.]*** (Moure's Bee Catalogue, 2008)

Johann Albert **Huber** (Born 10/7/1890 in Basel – Died 13/7/1972) Dr. Phil., and Biology Teacher, speciality in Lepidoptera. (Senckenberg Biographies, 2013)

iheringi, *Trigona* Ihering, H. (1903) - Named after the author H. **Ihering**; [*Nomen nudum* Ihering, 1903 attributed the name *Trigona* ***iheringi*** to Friese, but this name was never published by Friese]* (Moure's Bee Catalogue, 2008)

Hermann Friedrich Albrecht von **Ihering** (9 October 1850 – 24 February 1930) was a German-Brazilian zoologist. (Senckenberg Biographies, 2013)

jenningsi, *Melipona orbignyi* Cockerell, T. D. A. (1919) *[Typ. Loc.:* Las Cascadas, Canal Zone, Panama, collected by A. H. Jennings.]* (Moure's Bee Catalogue, 2008)

joearroyoi, *Geotrigona* Gonzalez & Engel, 2012 [Etym.: patronym for Alvaro José Arroyo González.]* (Moure's Bee Catalogue, 2008)

julianii, *Plebeia* Moure, 1962 - Named after the collector of *Plebeia (Plebeia)* ***julianii*** Moure, 1962 *Typ. Loc.:* Londrina, Paraná, Brazil, VIII.1960, L. **Juliani** leg.]* (Moure's Bee Catalogue, 2008)

junodi, *Trigona* Cockerell, 1932 Syn. n. = Named after the naturalists Henri-Alexandre **Junod** [Refers to Trigona togoensis **junodi** Cockerell: Cockerell 1919; Meliponiden **junodi** (Cockerell): Friese 1941; Axestotrigona **junodi** (Cockerell): Moure 1961]^ (Eardley, 2004)

Dr. Henri-Alexandre **Junod** (Born 1863/64 – Died 22/4/1934 in Geneva) was also a Missionary. (Senckenberg Biographies, 2013)

kinzelbachi *Liotrigona* Koch, 2010 – [dedicated to Prof. em. Dr Ragnar Kinzelbach]*# (*#Koch, 2010)

klossi, *Trigona (Hypotrigona) pendleburyi* variety Schwarz 1939 - Named after the collector [(C.B. **Kloss** and H. M. Pendlebury) of the N. Borneo. -Bettotan, nr. Sandakan, the specimen of this species .]# (Rasmussen, 2008)

Cecil Boden **Kloss** (1877–1949) was an English zoologist. He was an expert on the mammals and birds of Southeast Asia. (Wikipedia, Cecil Boden Kloss , 2013)

kohli, *Trigona* Friese, 1900 [*Etym.:* patronym for Custus F. F. **Kohl**.]* Also in relation with *Trigona* ***kohli*** *tarataensis* Schwarz, 1948. (Moure's Bee Catalogue, 2008)

Franz Friedrich Kohl (Born 13/1/1851 in St. Valentin, South Tyrol – Died 15/12/1924 in Traismauer) Prof. Middle School Bolzano and Innsbruck, Naturlh. Mus. Vienna. (Senckenberg Biographies, 2013)

kraussi, *Trigona (Geotrigona) acapulconis* Schwarz, H. F. (1951 [Typ. Loc.: Panama, Fort Clayton, XII.1946 collected by N. L. H. **Krauss** leg..]* (Moure's Bee Catalogue, 2008)

Noel Louis Hilmer Krauss (Born 1910 in Honolulu – Died 14/10/1996), Collection: Insects (Mexico and canal Zone) National Museum of Natural History, Washington. (Senckenberg Biographies, 2013)

langhofferi, *Trigona* Friese, 1900 - Named after the naturalists A. **Langhoffer** in relation to [Axestotrigona langhofferi (Friese)]^ (Eardley, 2004)

August **Long Hoffer** (Born 17/4/1861 – Died 28/3/1940) Professor of Zoology, Speciality in Various orders, Collection: Croatia Diptera in Zagreb Museum. (Senckenberg Biographies, 2013)

lehmanni, *Trigona* Friese, 1901; Relates to *Frieseomelitta* ***lehmanni*** (Friese, 1901) [Typ. Loc.: Popayan, Colombia Etym.: patronym for the Consul F. **Lehmann**.]* (Moure's Bee Catalogue, 2008)

lendliana, *Meliponula (Meliplebeia)* (Friese) - Named after the naturalists Adolf Lendl and relates to synonyms [*Trigona lendliana* Friese, 1900; Melipona (Trigona) lendliana (Friese); Plebeiella lendliana (Friese); Trigona (Meliplebeia) lendliana Friese; Meliplebeia lendliana (Friese); *Meliponula (Meliplebeia) lendliana* (Friese)]^ (Eardley, 2004) + -ana(see suffix **–ana**)

Lendl, Adolf 1862–1942 Professional insect collector, sold through companies or directly; material widely scattered. Insect dealer in Budapest about 1892/1913. - Own yield (esp. Col.) from Neuquen / Argentina: 1907 (Senckenberg Biographies, 2013)

lieftincki, *Sundatrigona* (Sakagami & Inoue, 1987) Named after the Zoologist Maurits A **Lieftinck** concerning [*Trigona (Trigonella) lieftincki* Sakagami & Inoue 1987][#] (Rasmussen, 2008)

Maurits Anne (Maus) **Lieftinck** (Born 18/9/1904 in Amsterdam – Died 13/4/1985 in Rhenen) was a Dutch biologist, director and zoologist. Lieftinck played in the Dutch East Indies an important role in entomology from 1947 to 1954 and was the most influential zoologist of Indonesia. (Wikipedia, Maurits Lieftinck, 2013) Portrait 20

llorentei, *Plebeia* Ayala, 1999 - Relates to [*Plebeia (Plebeia) llorentei* Ayala, 1999 *ETYM.:* patronym for Jorge E. **Llorente**]* (Moure's Bee Catalogue, 2008)

lucii, *Plebeia* Moure, 2004 [Etym: homage to Prof. Lucio A. de Oliveira Campos]* (Moure's Bee Catalogue, 2008)

lundelli,,*Paratrigona* (Schwarz, 1938) [Typ. Loc.: British Honduras [=Belize], December 1928, collected by C. L. **Lundell**.]* (Moure's Bee Catalogue, 2008)

lupitae, *Melipona (Melipona)* Ayala, 1999 [Etym.: patronym for Lupita Ayala**,** wife of R. Ayala of]* (Moure's Bee Catalogue, 2008) + ae (see suffix **–ae**)

lutzi, *Geotrigona* Camargo & Moure, 1996 [Etym.: patronym for Frank E. Lutz.]* (Moure's Bee Catalogue, 2008)

Frank Eugene Lutz (Born 15/9/1879 in Bloomsburg, Pennsylvania – Died 27/11/1943) was an American entomologist. In 1909, he became assistant curator of invertebrate zoology at American Museum, New York City, becoming associate curator in 1916 (Senckenberg Biographies, 2013)

magretti, *Trigona* Friese, 1900 - Named after the naturalist Paolo **Magrett** in relation also with [Hypotrigona **magretti** (Friese)and Trigona (Hypotrigona) **magretti** Friese]^ (Eardley, 2004)

Paolo Magrett (Born 15/12/1854 in Milano – Died 30/8/1913 in Cassina Amata, Milano) Assistant Professor at Univ. Inst Zool. Pavia. Collection: Hymenoptera (Africa: Sudan, Tunisia, Eritarea) to Mus. Civ. St. nav. Genova. (Senckenberg Biographies, 2013)

manni *Trigona* Cockerell (1912) - Named after the entomologist William Mann honouring the collector of [Trigona **manni** Cockerell, (1912). *Typ. Loc.:* Porto Velho, Rio Madeira, Brasil (**Mann** & Baker).]* (Moure's Bee Catalogue, 2008) see also ***mulfordi***

William M **Mann** (1886–1960) was a notable entomologist and the fifth director of the National Zoo in Washington from 1925-1956. He was assistant director of the Mulford Expedition to the Amazon River Basin, 1921-1922. (Wikipedia, William M. Mann, 2014)Portrait 35

margaritae, *Plebeia (Plebeia)* Moure, J. S. (1962). [Typ. Loc.: Pimenta Bueno, Rondônia, Brazil, collected by M. Alvarenga leg.. *Etym.:* patronym for Margarita, the wife of Colonel Moacyr Alvarenga.]* (Moure's Bee Catalogue, 2008)

martinezi, *Dolichotrigona* (Brèthes, 1920) - Named after the collector of [*Trigona **martinezi*** Brèthes, (1920) Typ. Loc.: Lima, Peru, XI.1919, by Ezequiel A. Martinez leg. [locality wrong, probably from Chanchamayo region according to Camargo & Pedro, 2005: 78].]* (Moure's Bee Catalogue, 2008)

matsumurai, *Trigona (Tetragona)* Sakagami 1959 [Typ. Loc.: SINGAPORE Sept. 22, 1932, S. Matsumura leg.][#] - Named after the entomologist Matsumura who collected this species. (Rasmussen, 2008)

Shōnen **Matsumura** (松村 松年 *Matsumura Shōnen*[?], (5/3/1872 – 7/11/1960) was a Japanese entomologist. Akashi, Hyōgo, Dr. Shōnen Matsumura established Japan's first course on entomology at Hokkaido University. The courses were both applied (on insects of importance in forestry and agriculture) and theoretical. (Wikipedia, Shonen Matsumura, 2014) Portrait 29

mayri, *Paratrigonoides* Camargo & Roubik, 2005 [*Etym.:* in honour of Ernst **Mayr**.]* (Moure's Bee Catalogue, 2008)

Ernst Walter **Mayr** (5/7/1904 – 3/2/2005) was one of the 20th century's leading evolutionary biologists. He was also a renowned taxonomist, tropical explorer, ornithologist, and historian. (Wikipedia, Ernst Mayr, 2014) Portrait 11

mazucatoi, *Trigona (Trigona)* Almeida, (1995) - Named after the technician - JMF Camargo **Menderson Mazucato** [see also ***mendersoni***]* (Moure's Bee Catalogue, 2008)

meadewaldoi, *Frieseomelitta* Cockerell, 1915 - Named after the entomologist Geoffrey **Meade-Waldo** and in relation with [*Trigona* Cockerell, 1915]* (Moure's Bee Catalogue, 2008)

Geoffrey **Meade-Waldo** (1884 – 1916) Hymenopterist at the Natural History Museum, London. (Senckenberg Biographies, 2013)

mendeli, *Trigona (Frieseomelitta)* Kerr, W. E. (1969). [*Nomen nudum*]* - Named after G.H. Mandel. (Moure's Bee Catalogue, 2008)

Gregor Johann **Mendel** (Born 22/7/1822 in Heinzendorf, Austrian Silesia Driven – Died 1/6/1884 in Brno) was an Austrian naturalist and an Abbot of the Augustinian Monastery in Brno. (Wikipedia, Gregor Mendel, 2014)

mendersoni, *Dolichotrigona* Camargo & Pedro, 2005 -Named after the collector and entomologist **Menderson Mazucato** [*Typ. Loc.:* Rio Ipixuna, Purus, Amazonas, Brasil, 20-23.I.1986, Camargo, Mazucato leg.. Etym.: patronym for Menderson Mazucato]* (Moure's Bee Catalogue, 2008) (see also ***mazucatoi***)

merrillae, *Melipona (Michmelia) seminigra* Cockerell, 1919 - Named after the naturalists named Merrill + -ae(see suffix **-ae**); [Refers to the collector, Harriet B. Merrill for Type species *Melipona pseudocentris* ***merrillae*** Cockerell, (1919)]* (Moure's Bee Catalogue, 2008) **Sp. 9**

Harriet B. Merrill was a female participant in The Stanford Expedition. (REHN, 1989)

Sp. 9 *Melipona seminigra merrilae* Photo: Marcos Ferreira

micheneri, *Melipona (Eomelipona)* Schwarz, 1951 - Named after the entomologist Charles Duncan Michener. Also relating to [*Melipona marginata* Schwarz, 1951, Etym.: patronym for Prof. C. D. **Michener.**]* (Moure's Bee Catalogue, 2008)

The entomologist American Charles Michener was born in Pasadena, California, on 22/9/1918. Much of his career has been devoted to the systematics and natural history of bees. (Wikipedia, Charles Duncan Michener, 2014)

mohandasi, *Lisotrigona* Jobiraj & Narendran 2004: [Typ. Loc.: India, Kerala Forest Research Institute (Peechi), collected by K. Mohandas,][#] (Rasmussen, 2008)

moratoi,*Dolichotrigona* Camargo & Pedro, 2005 [Typ. Loc.: Brasil, Acre, Parque Nacional Serra do Divisor collected by E. F. **Morato** leg. Etym.: patronym for Dr. Elder Ferreira **Morato**]* (Moure's Bee Catalogue, 2008)

Dr. Elder Ferreira Morato from Universidade Federal do Acre – specialise in Plants & animal science.

moorei*, Sundatrigona* (Schwarz, 1937) - Named after Frederick Moore and also in relation with [*Trigona moorei* **Schwarz** 1937][#] (Rasmussen, 2008)

Frederic **Moore** FZS (13/5/1830 – 10/5/1907) was a British entomologist. Moore was an associate of the Linnean, a member of the Entomological Society of London, a corresponding Entomological and the Entomological Society of the Netherlands. (Senckenberg Biographies, 2013)

mourei*, Partamona* Camargo, 1980 & Schwarziana **mourei** Melo, 2003 [*Etym.:* patronym for Dr. Jesus Santiago Moure]* (Moure's Bee Catalogue, 2008)

Professor Padre Jesus Santiago Moure, a Brazilian. Born on 2/11/1912 in the city of Ribeirão Preto (São Paulo) – Died 10/6/2010, was a founder of the Faculty of Philosophy, Sciences and Letters of the Federal University of Paraná, as professor of the chair of Zoology, Programs Postgraduate Entomology and Zoology and Entomology Collection that bears his name. (Wikia, Jesús Santiago Moure) Portrait 11

moureana*, Plebeia* Ayala, 1999 [*Etym.:* patronym for Padre Jesús Santiago **Moure**.]* (Moure's Bee Catalogue, 2008) + ana(see suffix **–ana**) see also ***mourei.*** (Databank, Padre Jesus Santiago Moure, 2013)

mocsaryi*, Trigona* mocsaryi Friese, H. (1900). [*Etym.:* patronym for the curator Alexander Mocsáry.]* (Moure's Bee Catalogue, 2008)

muelleri*, Leurotrigona* Friese, 1900 and also in relation with *Trigona* ***muelleri*** Friese, H. (1900). [Etym.: patronym for Prof. Herm. **Mueller**.]* (Moure's Bee Catalogue, 2008)

Heinrich Ludwig Hermann Müller (1829–1883) was a German botanist who provided important evidence for Darwin's evolution theory. Müller was an early investigator of coevolution. (Wikipedia, Hermann Muller, 2014) Portrait 35

mulfordi*, Oxytrigona* Schwarz, 1948 - Named after the **Mulford** Expedition of the Amazon Basin (1921-1922). *Trigona* Schwarz, 1948 [*Typ. Loc.:* Tumupasa, Bolivia, XII.1921-1922 collected by W. M. Mann leg..]* (Moure's Bee Catalogue, 2008)

Collector of this Type of species, Willliam M. Mann, was assistant Director of the **Mulford** Biological Exploration of the Amazon Basin (1921-1922). The trip was reported to have been the most ambitious and well equipped scientific expedition ever to set sail for South America at that time. The trip was financed by the Mulford Pharmaceutical Company (Philadelphia) (Hutchison, 1921)

nataliae, *Hypotrigona* Moure, 1950 – This name was dedicated to the wife of the collector of [*Trigonisca* ***nataliae*** (Moure, 1950) *Typ. Loc.:* Imperatriz, Maranhão, Brazil, 5-11.VIII.1949, by T. Dobzhansky leg.. *Etym.:* matronym for Mrs. **Natalie** P. Sivertzev-Dobzhansky.]* (Moure's Bee Catalogue, 2008)

neavei, Meliplebeia beccarii (Cockerell) - Named after the naturalists Sheffield Airey **Neave,** and also relates to [*Trigona (Meliponula) africana* ***neavei*** Cockerell, 1934 Syn.n.**]**^ (Eardley, 2004)

Sheffield Airey **Neave** CMG OBE MRCP[1] (20 April 1879 - 31 December 1961) was a British naturalist and entomologist. (Wikipedia, Sheffield Airey Neave, 2013)

nilssoni, *Liotrigona* Michener, 1990 - Named after the entomologist Anders N. Nilsson [The scientific name of the species was first published in 1990 by valid Michener.]^* (Ruggiero, 2008)

Anders N. Nilsson, Associate Professor in Zoology at the Department of Biology, University of Umea, Sweden. He is the editor of a taxonomic handbook of the aquatic insects of N.W. Europe. Mogens Holmen, a scientific candidate (1968) in Biology, University of Copenhagen, is an amateur entomologist, professionally employed at the Landscape Department of the Frederiksborg County Council. (Nilsson, 2014)

ogilviei, *Melipona* Schwarz, 1932 and concerning the collector of the Type species [*Melipona puncticollis* Schwarz, 1932 Typ. Loc.: Marshall Fall, Mazaruni River, British Guiana, 8.IV.1929 by J. Ogilvie. Etym.: named in honour of Mr. J. Ogilvie.]* (Moure's Bee Catalogue, 2008)

okazawai, *Trigona (Pacificotrigona)* Tadauchi *et al.* 1998: Named after the entomologist Takao Okazawa and also in relation to *[Trigona (Pacificotrigona)* ***okazawai*** *irianensis* Tadauchi *et al.* 1998][#] (Rasmussen, 2008)

onorei, *Paratrigona* Camargo & Moure, 1994 **[***Etym.:* dedicated to Prof. Giovane Onore.]* (Moure's Bee Catalogue, 2008)

Bro. Giovanni **Onore,** an Italian scientist who has lived in Ecuador for 25 years, is the former director of the Museum of Invertebrate Zoology at the Pontifical Catholic University of Ecuador Quito. His areas of study include biology, phytopathology, entomology, invertebrate zoology, arthropods, highland insects, insect pests and bees. He also has significant knowledge of nature conservation with a special emphasis on the Amazonian region. (Marianist, 2014) Portrait 20

orbignyi *Trigona* Guérin-Méneville, (1844). [Typ. Loc.: Chiquitos, Bolívia probably collected by A. d'Orbigny (July 1831)]* (Moure's Bee Catalogue, 2008)

Alcide Dessalines **d'Orbigny** (6/9/1802 – 30/6/1857) was a French naturalist who made major contributions in many areas, including zoology (including malacology), palaeontology, geology, archaeology and anthropology. (Senckenberg Biographies, 2013)

pagdeni, *Tetragonula* (Schwarz, 1939) - Named after the naturalists H T **Pagden** (speciality Hymenoptera) and also in relation with *Trigona (Tetragona) fusco-balteata* variety ***pagdeni*** Schwarz 1939 (Rasmussen, 2008) Sp. 10

Sp. 10 *Tetragonula pagdeni* Photo: Joe Teamworkshop.

pagdeniformis, *Tetragonula* (Sakagami, 1978) - relating to the form of *pagdeni*. See also ***pagdeni***.

parkeri, *Plebeia* Ayala, 1999 [*Etym.*: patronym for Dr. Frank D. Parker.]* (Moure's Bee Catalogue, 2008)

pavani, *Duckeola* Moure, 1963 - Named after the collector of *Trigona* Moure, 1963 [*Typ. Loc.*: Içana, Amazonas, Brasil, 25.IX.1952 by T. Dobzhansky & C. **Pavan** leg..]* (Moure's Bee Catalogue, 2008)

Crodowaldo **Pavan** (Portuguese pronunciation: [krodoˈvałdu paˈvẽ]; (1/12/1919 – 3/4/2009) was a Brazilian biologist and geneticist, and a scientific leader in Brazil. (Wikipedia, Crodowaldo Pavan, 2013)

pearsoni, *Partamona* Schwarz, 1938 Named in honour of Nathan Everett Pearson (a member of the Mulford expedition in the Amazon (Hutchison, 1921)), also relates to [***Trigona*** Schwarz, 1938]* (Moure's Bee Catalogue, 2008)

peckolti, *Partamona* (Friese, 1901) [*Etym.*: patronym for Dr. Theodor **Peckolt**.]* (Moure's Bee Catalogue, 2008) Sp. 11

Theodor **Peckolt** (1822–1912) was a German-born naturalist, botanist, phytochemist and pharmacist who worked in Brazil. (Wikipedia, Theodor Peckolt, 2013)

Sp. 11 *Partamona peckolti* (Wikipedia, Partamona peckolti 2014)

pendleburyi, *Pariotrigona* (Schwarz, 1939) also relates to [*Trigona (Hypotrigona) pendleburyi* Schwarz 1939 Typ. Loc.: Malaysia Cameron Highlands, Rhododendron Hill, collected by H. M. Pendlebury][#] (Rasmussen, 2008)

Henry Maurice **Pendlebury**. (1893-1945) was a systematic entomologist and co-author of The Butterflies of Peninsula Malaysia book. The Pendlebury Collection of beetles, collected from 1914 to 1939, is housed in the Museum of Zoology, University of Malaya. (Gazi, 2013)

pfeifferi, *Trigona* Friese 1925 Named after the naturalists Ida Laura **Pfieffer** relates to [*Trigona pfeifferi* Friese 1925][#] (Rasmussen, 2008) Portrait 29

pseudograeffei, *Trigonisca* Albuquerque, P. M. C. (1992). (*Nomen nudum*) = pseudo (Latin prefix 'pseudo-' False.) + graeffei (see graeffei)

portoi, *Frieseomelitta* Friese, 1900 also relates to *Trigona* Friese, 1900 [Etym.: patronym for Dr. Porto, sub-director of the Museum Paraense Emílio Goeldi.]* (Moure's Bee Catalogue, 2008)

pothieri, *Hypotrigona* Lobreau-Callen et al. 1990 [nomen nudum - Named after the Entomologist Pothier] (Eardley & Urban, 2010)

reepeni, *Tetragonula* (Friese, 1918) - Named after the naturalists Buttel-**Reepen,** who collected the type series of *Trigona* ***reepeni*** **Friese** 1918. (Rasmussen, 2008) See also ***buttelli***

rhumbleri, *Trigona* Friese, H. (1900) [*Etym.:* patronym for Prof. L. **Rhumbler**.]* (Moure's Bee Catalogue, 2008)

Prof. Dr. Ludwig Rhumbler (Born 03/07/1864 - Died 06/07/1939 in Hann, Munden) Speciality in Entomology (Senckenberg Biographies, 2013)

richardsi, *Melipomila (Axestotrigona)* Darchen comb. n. - Named after the naturalists Owain Westmacott **Richards** also in relation with [*Trigona (Axestotrigona)* ***richardsi*** Darchen, 1981]^ (type details and depository unknown). (Eardley, 2004)

Owain Westmacott **Richards** FRS (31/12/1901 – 9/11/1984) was a British entomologist. He was Professor of Zoology and Applied Entomology, Imperial College, London, 1953–1967 (Wikipedia, Owain Richards, 2013)

rinconi, *Paratrigona* Camargo & Moure, 1994 [*Etym.:* patronym for Prof. José **Rincon**.]* (Moure's Bee Catalogue, 2008)

rohwerae, *Trigona silvestriana* Cockerell, (1925) [*TYP. LOC.:* Trinidad, B. W. I., 7 and 16.IV.1923, collected by Hester M. Rohwer leg..]* (Moure's Bee Catalogue, 2008) + ae (see suffix **–ae**)

rondoni, *Dolichotrigona* Camargo & Pedro, 2005 [*Typ. Loc.:* Brasil, **Rondônia**, Nova Mamoré *Etym.:* patronym for Marechal Cândido Mariano da Silva **Rondon**.]* (Moure's Bee Catalogue, 2008)

Cândido Mariano da Silva **Rondon** or **Marechal Rondon** (5/5/1865 – 19/1/1958) was a Brazilian military officer who is most famous for his exploration of Mato Grosso and the Western Amazon Basin and his lifelong support of Brazilian indigenous populations. He was the first director of Brazil's Indian Protection Bureau (SPI/FUNAI) and responsible for the creation of the Xingu (Wikipedia, Cândido Rondon, 2014) Portrait 35

roubiki, *Trigonisca* Albuquerque & Camargo, 2007 [*Typ. Loc.:* Chepo, 25 Km NE, Panamá Prov., Panamá, 27.X.1983, D. **Roubik** leg.. *Etym.:* patronym for David W. **Roubik.**]* also relates to *Meliponula (Meliplebeia)* ***roubiki*** *sp.n.*, [Etym.: This species was recognized as being new by David **Roubik** and is named for him.]^ (Eardley, 2004)

David Ward Roubik, a resident bee expert at the Smithsonian Tropical Research Institute; (NIEH, 2000) Portrait 20

rozeni, *Liotrigonopsis* Engel[&&] Etym.: The specific epithet is a patronymic honouring Dr. Jerome G. Rozen, Jr. (Moure J.S. & Camargo J.M.F., 1978)

ruspolii, *Hypotrigona* (Magretti), - Named after Eugenio **Ruspoli** (1866–1893) and also in relation with Trigona ruspolii Magretti, 1898 (Eardley & Urban, 2010)

Prince Eugenio Ruspoli (1866-1893), an Italian explorer, gave his name to one of the world's most beautiful and rare birds, the Ethiopian Prince Ruspoli's Turaco, trampled to death by an angry elephant. (BEARMAN, 2011) Portrait 20

sakagamii, *Trigona*Moure, [in Kerr et al., 1967 *Nomen nudum*]* - Named after Prof. Shoichi Francisco Sakagami (Moure's Bee Catalogue, 2008)

S. F. Sakagami, Emeritus Professor of the Hokkaido University, was born (4/1/1927) in Ichikawa (Central Japan) and passed away (4/11/1996) in Sapporo (North Japan). Prof. Sakagami was a very prolific scientist, as he authored over 320 papers and eight books, of which concentrated on the Hymenoptera. (Kevan P & Imperatriz Fonseca VL (eds), 2002) Portrait 35

salti, *Melipona* Schwarz, 1932 [Etym. Refers to George Salt]* (Moure's Bee Catalogue, 2008)

santhilarii, *Melipona*Lepeletier, 1836 [Typ. Loc.:Rio de Janeiro, Brazil, Col, **Saint Hilaire** and also relates to *Melipona santhilarii paulomajor* Lepeletier, 1836]* (Moure's Bee Catalogue, 2008)

Étienne Geoffroy **Saint-Hilaire** (15/4/1772 – 19/6/1844) was a French naturalist who established the principle of "unity of composition". He was a colleague of Jean-Baptiste Lamarck and expanded and defended Lamarck's evolutionary theories. (Wikipedia, Étienne Geoffroy Saint-Hilaire, 2014)

sawadogoi *Meliponula* (Axestotrigona) Darchen comb. n. - Named after the African naturalists Sawadogo and also [*Trigona (Axestotrigona) sawadogoi* Darchen, 1970 (material depository unknown)]^ (Eardley, 2004)

schencki, *Melipona*Gribodo, 1893 [Etym.: patronym for Dottore Schenck.]* (Moure's Bee Catalogue, 2008)

schmidti, Dactylurina (Stadelmann), Type material collected by C W **Schmidt**, also in relation with [*Melipona* ***schmidti*** Stadelmann, 1895 and *Trigona* ***schmidti*** (Stadelmann): Friese 1909]^ (Eardley, 2004)

Carl Wilhelm **Schmidt** (1859-1924) German entomologist (Senckenberg Biographies, 2013)

schrottkyi, *Friesella* Friese, 1900 - Named after the naturalists Kurt Schrottky, the collector of the [Type species and also concerning *Trigona* ***schrottkyi*** Friese, 1900]* (Moure's Bee Catalogue, 2008) Sp. 12

Sp. 12 *Friesella schrottkyi* Photo: via Maria Rosa Bernasconi

Schrottky, Kurt [Curt] (1874–1937): Curt Schrottky was one of the early European entomologists to reside in South America. Many of his primary types (syntypes and holotypes) were originally in his private collection, partially destroyed, and the remaining part scattered among multiple institutions. Other specimens were exchanged before the destruction of his main collection to institutions in South America, the USA, and Europe. (CLAUS RASMUSSEN et al., 2009) Portrait 20

schulthessi, *Dolichotrigona* Friese, 1900 also relates to *Trigona* ***schulthessi*** Friese, 1900 [Etym.: patronym for Dr. Ant. v. Schulthess.]* (Moure's Bee Catalogue, 2008)

Anton von Schulthess-Schindler Rechenberg (Born 14/1/1855 – Died 7/11/1941 in Zurich) Entomologist Speciality in various orders. (Senckenberg Biographies, 2013)

schultzei, *Nannotrigona* (Friese, 1901) *Etym.:* patronym for Dr. Leo **Schultze**.]* (Moure's Bee Catalogue, 2008)

Dr. Leonhard Schultze (Born 28/5/1872 in Jena – Died 29/3/1955 in Marburg (Lahn)) (Senckenberg Biographies, 2013)

schwarzi, *Melipona (Eomelipona)* Moure, 1963 - Named after the naturalists Herbert F. **Schwarz** also [*Melipona schwarzi* Moure, (1963)]*

Herbert Ferlando **Schwarz** (07/09/1883 – 02/10/1960) was a toy dealer and part-time amateur entomologist specializing in Hymenoptera. (Senckenberg Biographies, 2013)

silvestrii, *Trigona* Friese, H. (1902). syn. to *Frieseomelitta silvestrii* (Friese, 1902) [Worker specimen collected in Coxipó, Mato Grosso, Brazil, 1900, by Filippo **Silvestri** leg..]* (Moure's Bee Catalogue, 2008) Sp. 13

Sp. 13 *Friesomelitta silvestrii* Photo: via Mari Rosa bernasconi

Filippo **Silvestri** (Born 22/6/1873 in Bevagna - Died 10/6/1949 in Portici) was an Italian entomologist. He specialised in world Protura, Thysanura, Dipluraand Isoptera, and worked on Hymenoptera, Myriapoda and Italian Diptera. (Wikipedia, Filippo Silvestri, 2013)

simpsoni, *Meliponula* (Axestotrigona) Moure comb.n. Named after G. Simpson and also relates to [Axestotrigona simpsoni Moure, 1961]^ (Eardley, 2004)

George Gaylord **Simpson** (16/6/1902 – 6/10/1984) was an American palaeontologist. Simpson was perhaps the most influential palaeontologist of the twentieth century, and a major participant in the modern evolutionary synthesis (Wikipedia, George Gaylord Simpson, 2014)

sirindhornae, *Tetragonula* (Michener & Boongird, 2004) [Etym.: This species was dedicated to HRH Princess Sirindhorn of Thailand on the occasion of her 48th birthday celebrations]`` (Boongird, 2004)/ Portrait 29

smithii, *Melipona* Bingham 1897 [Unnecessary replacement name for *Trigona ruficornis* Smith][#] - Named after the entomologist F. Smith

Frederick **Smith** - (Born 30/12/1805 – Died 16/2/1879) was a British entomologist. Smith worked in the zoology department of the British Museum from 1849, specialising in the Hymenoptera. In 1875 he was promoted to Assistant Keeper of Zoology. His publications included *Catalogue of Hymenopterous Insects* (7 parts, 1853–1859) and parts 5 (1851) and 6 (1852) of the *Nomenclature of Coleopterous Insects*. (Wikipedia, Frederick Smith, 2013)

staudingeri, *Dactylurina* (Gribodo), - Named after the naturalist Otto Staudinger also relating to [*Trigona* ***staudingeri*** Gribodo, 1893]^ (Eardley, 2004)

Otto **Staudinger** (2/5/1830 – 13/10/1900) was a German entomologist, and a natural history dealer considered one of the largest in the world specialising in the collection and sale of insects to museums, scientific institutions, and individuals (Wikipedia, Otto Staudinger, 2014) Portrait 29

symei, *Austroplebeia* (Rayment, 1932) Named after the naturalists Syme also concerning [*Trigona symei* Rayment 1932][#] (Rasmussen, 2008)

tavaresi, *Dolichotrigona* Camargo & Pedro, 2005 [Etym.: patronym for José Amílcar Tavares Filho.]* (Moure's Bee Catalogue, 2008)

townsendi, *Trigonisca* (Cockerell, 1911) - Named after Prof Charles H. T. Townsend, also the collector of *Trigona* ***townsendi*** Cockerell, (1911)[*Typ. Loc.:* Piura, Peru, October, by Townsend]* (Moure's Bee Catalogue, 2008)

Charles Henry Tyler **Townsend** (Born 5/12/1863 in Oberlin – Died 17/3/1944 in Itaquaquecetuba near São Paulo) was an American entomologist and biologist. From 1919 he was chief entomologist of the state of São Paulo, Brazil. (Wikipedia, Charles Henry Tyler Townsend, 2013) Portrait 20

Sp. 14 *Tetragonula valdezi* Photo: Mohd Zulkarnain Abdul Rahman

valdezi, *Tetragonula* (Cockerell, 1918) - Named after the naturalists Samuel Macias Valdez also relates to [*Trigona* ***valdezi*** Cockerell 1918 by original description]# (Rasmussen, 2008) Sp. 14

virgilii, *Trigona capitata* Friese, H. (1901) (junior subjective synonym); [Typ. Loc.: Blumenau, Santa Catarina, Brasil, Oct. 1898, collected by G. Virgil leg..]* (Moure's Bee Catalogue, 2008)

vitae, *Partamona* Pedro & Camargo, 2003 [*Etym.:* patronym for Patricia **Vit**.]* (Moure's Bee Catalogue, 2008) + -ae (see suffix **–ae)** Portrait 29

voeltzkovi, Liotrigona Friese 1900. - Named after the naturalists Alfred Voeltzkov. [The scientific name of the species was first published in 1900 by Frisian valid.] ^* (ITIS, 2014)

wallacei, *Trigona wallacei* Tadauchi ***et al.*** 1998: *Nomen nudum*. - Named after Alfred R. Wallace ; [The name is unavailable for taxonomic purposes, as no description or definition of the taxon was provided (ICZN, Art. 15.1)]# (Rasmussen, 2008)

Alfred Russel **Wallace** OM FRS (8/1/1823 – 7/11/1913) was a British naturalist, explorer, geographer, anthropologist and biologist. He is best known for independently conceiving the theory of evolution through natural selection; his paper on the subject was jointly published with some of Charles Darwin's writings in 1858. (Wikipedia, Alfred_Russel_Wallace, 2014)Portrait 20

wasbaueri, *Paratrigona* Gonzalez & Griswold, 2011 *Etym.:* the specific epithet is a patronym honouring Marius **Wasbauer**]* (Moure's Bee Catalogue, 2008)

websteri, *Austroplebeia* (Rayment, 1932) - Named after Francis Marion Webster and also relates with [*Trigona* ***websteri*** Rayment 1932 by original description][#] (Rasmussen, 2008)

Francis Marion Webster (Born 02/08/1849 in Lebanon, NH - Died 01/02/1916 in Columbus, OH) (Senckenberg Biographies, 2013)

williana, *Trigona* Friese, 1900 [*Etym.:* patronym for Prof. L. Will.]* (Moure's Bee Catalogue, 2008) + -iana (see suffix -iana)

wittmanni, *Plebeia* Moure & Camargo, 1989 [*Typ. Loc.:* Canguçu, Santo Amor, Rio Grande do Sul, Brazil, 1987, collected by D. **Wittmann** leg..]* (Moure's Bee Catalogue, 2008)

weyrauchi, *Tetragonisca* (Schwarz, 1943) **[***Typ. Loc.:* Valle Chanchamayo, Peru, 800 mts., 1939, Col. W. **Weyrauch**.]* (Moure's Bee Catalogue, 2008)

Wolfgang Karl Weyrauch(1907-1970) (Senckenberg Biographies, 2013)

wheeleri, *Scaptotrigona* (Cockerell, 1913) [*Typ. Loc.:* Escuintla, Guatemala, 20.XII.1911, collected by William M. **Wheeler** leg.]* (Moure's Bee Catalogue, 2008)

ziegleri *Tetragona* (Friese, 1900) [*Etym.:* patronym for Prof. H. E. Ziegler (1858-1925).]* also relates to *Trigona* ***ziegleri*** *mayarum* Cockerell, (1912) (Moure's Bee Catalogue, 2008)

Sp. 15 *Tetragonula zucchii* Photo Acai Ibrahim

zucchii, *Tetragonula* (Sakagami, 1978) – Named after the naturalists Prof. Dr. Ronaldo Zucchi and also relates to [*Trigona (Tetragonula)* ***zucchii*** Sakagami 1978][#] (Rasmussen, 2008) Sp. 15

Prof. Dr. Ronaldo Zucchi was a Director for the University of Sao Paulo, Brazil (1984-1988) (Zucchi, 2009) Portrait 29

Legend for Patronymics section:

[..]* - Online Moure.s catalogue for the Neotropical bee species. http://moure.cria.org.br/catalogue

[..]# - Catalog of the Indo-Malayan/Australasian stingless bees (Hymenoptera: Apidae: Meliponini)

CLAUS RASMUSSEN (2008)

[..]^ - Taxonomic revision of the African stingless bees(Apoidea: Apidae: Apinae: Meliponini) C D Eardley (2004)

[..]``- A New Species of *Trigona* from Peninsular Thailand (Hymenoptera: Apidae: Meliponini) Charles D. Michener and Somnuk Boongird (2004)

[..]^* - (2008) Integrated Taxonomic Information System (ITIS) Bee Checklist October 3, 2008, http://www.itis.gov/beechecklist.html

[..] *#Combining morphology and DNA barcoding resolves the taxonomy- KOCH, H. 2010:

[1] - Personal comments/notes from Dr. Claus Rasmussen

[2] - Catalogue of Afrotropical bees (Hymenoptera: Apoidea: Apiformes) - Connal Eardley & Rosalind Urban

Gallery - Portraits of Patronyms

Portrait 11 Portrait of Heinrich Friese (1860–1948). (Claus Rasmussen & John S. Ascher, 2008)

Portrait 11 Atahualpa *Sapa Inca (14th)* Portrait drawn by employees of Pizarro, 1533 (Wikipedia, Atahualpa, 2014)

Portrait 11 A photograph of Henry Walter Bates. (1825-1892) (Wikipedia, Henry Walter Bates, 2014)

Portrait 11 Edward Charles Stuart Baker Created: January 1, 1933 (Wikipedia, E. C. Stuart_Baker, 2014)

Portrait 11 Created: January 1, 1908 (Wikipedia, Charles Thomas Bingham, 2013)

Portrait 11 William Beebe on an expedition to Guiana in 1917. (Wikipedia, William Beebe, 2014)

Portrait 11 Frederick William Beechey, portrait by George Duncan Beechey, painted circa 1822. (Wikipedia, Frederick William Beechey, 2014)

Portrait 11 Prof Jesus Santiago Moure (Moure's Bee Catalogue, 2008)

Portrait 11 Ernst Mayr in 1994 (Wikipedia, Ernst Mayr, 2014)

Portrait 20 David Ward Roubik Photo: Eric Tourneret.

Portrait 20 Charles Henry Tyler Townsend (NMSU 2013)

Portrait 20 Kurt Schrottky (Rasmussen 2009)

Portrait 20 Prince Eugenio Ruspoli (Conniff 2009)

Portrait 20 Peter Davenport Photo by Anne Dollin

Portrait 20 Giovanni Onore Photo by: Patricia Vit

Portrait 20 Maurits Lieftinck (Wikipedia, Maurits Lieftinck, 2013)

Portrait 20 Prof. João Maria Franco de Camargo

Portrait 20 Alfred Russel Wallace (Wikipedia 2014)

Portrait 29 Shonen Matsumura Created: November 3, 1954, (Wikipedia, Shonen Matsumura, 2014)

Portrait 29 Otto Staudinger 1830-1900 (Wikipedia, Otto Staudinger, 2014)

Portrait 29Theodore Dru Alison Cockerell (1866–1948) (Wikipedia, Theodore Dru Alison Cockerell, 2013)

Portrait 29 Ida Pfeiffer dressed for a collection foray with an insect net and a specimen container slung over her shoulder. (Wikipedia, Ida Laura Pfeiffer, 2014)

Portrait 29 Goeldi Emilio Augusto 1859-1917 (Wikipedia, Emil Goeldi, 2014)

Portrait 29 Prof Dr Patricia Vit (Pers. Comm. 2015)

Portrait 29 Dr. Ronaldo Zucchi

Portrait 29 HRH Princess Maha Chakri Sirindhorn 2010 (Wikimedia, 2014)

Portrait 29 Warwick Kerr 1922 (Wikia, Warwick Kerr)

Portrait 35 Paolo Nogueira-Neto (Nogueira-Neto, 1997)

Portrait 35 Shoichi Francisco Sakagami (Kevan P & Imperatriz Fonseca VL (eds), 2002)

Portrait 35 A young Cândido Rondon in the Amazon Created: January 1, 1891 (Wikipedia, Candido Rondon, 2014)

Portrait 35 Theodosius Dobzhanskyi (Wikipedia, Theodosius Dobzhansky, 2014)

Portrait 35 Prof. Dr. Hermann Müller (Lippstadt) (Wikipedia, Hermann Muller, 2014)

Portrait 35 William M. Mann photo by Lucille Mann, 1939 (Wikipedia, William M. Mann, 2014)

Chapter 5

Peoples and Places

In taxonomy (the process or system of describing how different living things are related by putting them in groups) or studying the general principles of scientific classification (also known as systematics), ethnicity and locality can sometimes influence the naming process. It can even be named after a region, community, municipality or country.

Ethnicity:

It concerns large groups of people classed according to the common racial, national, tribal, religious, linguistic, or cultural origin or background.

Associated with or belonging to a particular race or group of people who have a different culture from the main culture. Comprise of the following:-

Ethnonym - From *the-(alternative to ethno)* + *-onym* (From Ancient Greek ὄνυμα (*onuma*), Doric dialectal form of ὄνομα (*onoma*, “name”).) **ethnonym** (*plural* **ethnonyms**) - The name of an ethnic group. e.g. a*zteca*

Endonym - *From endo-* ("inside") + *-onym* ("name") - **endonym** (*plural* **endonyms**) - The name used by a group or category of people to refer to themselves or their language, as opposed to a name given to them by other groups. e.g. *minangkabau*

Exonym From *exo-* (“outside”) + *-onym* (“name”) - **exonym** (*plural* **exonyms**) - A name given to a group or category of people by a secondary person or persons other than the people it refers to, e.g. *pipioli*

Xenonym - *From xeno-* ("foreign") + *-onym* ("name") **xenonym** (*plural* **xenonyms**) - A name for a people or a language or a city etc. which is not used by the natives themselves. e.g. *paraense*

Demonym – From *demo-* Ancient Greek δῆμος (*dēmos*, “people”) + *-onym* ὄνυμα (*onuma*, “name”). - **demonym** (*plural* **demonyms**) The name for an inhabitant or native of a specific place that is derived from the name of the place. e.g. *peruviana*

Theonym - A **theonym** (from classical Greek *theos* "god" and *-onym* "name") is a proper name of a deity, e.g. *amalthea*

Locality:

A particular place or area. May also depict the bearing concerning a particular location (Northern, Southern, Eastern or Western.) Other references to places may indicate whether arboreal, terrestrial or below ground (subterranean).

Alternative forms

- place-name
- place name

From *place* + *name*. **placename** (*plural* **placenames**) - The name of a place; a word or phrase which indicates a particular location or position.

How to use this Section

The purpose of this section is to show the connections between Taxa and peoples and places. It has been compiled and sorted from many different works and catalogues. Various styles and font formats have been used to differentiate the works and catalogues being referenced.

Sample entry

species or subspecies taxon (bold italics font)

variety, subspecies or trinomial name when exist (italics font)

Alternative combinations, synonymy, trinomial name or others species.(italics font)

genus and/or subgenus (italics font)

Author Surname with or without year named.

Taxon root word deratives and / or definition (blue font)

aequatoriana, *Partamona* Camargo, 1980 also relates to [*Melipona* Schulz, 1904 Typ. Loc. Esmeraldas, **Ecuador**.]* = aequator English æquator (*plural* **æquators**) - Archaic spelling of **equator** (**equator** (*plural* equators) - (often "the Equator") An imaginary great circle around the Earth, equidistant from the two poles, and dividing earth's surface into the northern and southern hemisphere.) + -iana (see suffix **–iana**) **Ecuador** is a small country in South America. It is along the equator. **Ecuador** officially the **Republic of Ecuador** (Spanish: *República del Ecuador* [reˈpuβlika ðel ekwaˈðor], which translates as "Republic of the Equator")

Terminates or suffixes when used to form the epithets in bracketed blue bold font..

Additional notes and/or trivia regarding epithet in (reduced grey font)

Date /year the Type specimen was examined or collected in numeral formats as in Citation.

Type Location with pertaining **location** in **bold font**

Etymology when provided by the cited literature in normal black font

manauara, *Celetrigona* Camargo & Pedro,. (2009). Typ. loc.: **Manaus**, Boulevard, AM, Brasil, 1239, ninho 1, 14.IV.1986, M. V. B. Garcia.; Etym.: Manauara is the common denomination to the people born in Manaus, Amazonas, Brazil, the type locality of the species here described.]* = Refers to **Manaus**, + ara (**āra** f (*genitive* **ārae**); *first declension* - altar, sanctuary, refuge); Manaus a **city** in **Brazil**, the **capital** of the state of **Amazonas**.

Symbol marker of square brackets corresponding to legend or citation key.

Name of specimen collector (leg.) or owner of the specimen collection (coll.) may also be Name of collecting expedition.

Meliponine Species Taxon Depicting Ethnicity or Locality

The epithets are listed below in alphabetical order by species taxa. The suffixes and terminates used in the species taxa are defined at the end of this list. A Tupi language Glossary is appended in the final pages.

Suffixes and terminates used in this section:

-aca = Spanish acá - *(Used with the verb of motion)* **hither**, towards the speaker; **here**; Serbo-Croatian - Suffix **-ača** (*Cyrillic spelling* -ача) - Suffix appended to words to create a feminine noun, usually denoting a possession, or place where the action occurs

-ae = suffix -ae Representing both nominative and vocative plural case endings belonging to Latin first-declension feminine nouns. English ae Variant of æ. - **æ** - A ligature of vowels a and e. Usage notes; Normally used for words of either Ancient Greek or Latin origin.;

-ana = From Latin suffix -ānus m (feminine -āna, neuter -ānum); first/second declension - Of or pertaining to; Alt. form -ian; usually indicates a relationship of position, possession, or origin.

-ara = ara (**āra** f (genitive **ārae**); first declension - altar, sanctuary, refuge);

-aris = Latin Suffix **-āris** m, f (neuter -āre); the third declension - Used to form an adjective, usually from a noun, indicating a relationship or a 'pertaining to'. Usage notes: The suffix -āris is added to a noun to form an adjective of relationship to that noun.

-alis = Latin Suffix **-ālis** m, f (neuter -āle); third declension - Used to form adjectives of relationship from nouns or numerals. Usage notes: The suffix *-ālis* is added to a noun or numeral to form an adjective of relationship to that noun

-cola = Latin cōlā - a first-person singular present active imperative of cōlō - present active colō, present infinitive colere, perfect active coluī, supine cultum - till, cultivate., **inhabit.;** strain.

-ense = Borrowed from Latin *-ēnsis* ("originating in"). Spanish Suffix **-ense -** Used to create adjectives and nouns that denote ethnonyms,

-ensis = Latin Suffix -ēnsis m, f (neuter -ēnse); third declension - Of or **from** a place.

-ia = From Latin *-ia* and Ancient Greek -ία (*-ia*), -εια (*-ia*), which form abstract nouns of feminine gender. The suffix **–ia** Used in forming names of **countries**, diseases, flowers, Synonyms (*collections of things*): -ana, -ica

-ica = From Latin suffix **-icus** m (*feminine* **-ica**, *neuter* **-icum**); *first/second declension* - Belonging to, Derived from, Of or pertaining to; connected with; -ish.

-nis = Old English suffix **-nis** f Used to form abstract nouns, usually from adjectives, denoting quality or state.

Ethnicity & Locality Epithets:

abunensis, *Melipona* Cockerell also relates to [Melipona abunensis Cockerell, 1912. *Typ. Loc.:* Abuná, Rio Madeira, Rondônia, Brazil]* (Moure's Bee Catalogue, 2008) = pertaining to Abuna + -ensis (see suffix –**ensis**);

Abuná can refer to Abuna River, the river between Bolivia and Brazil and Abuná Province in Brazil.

Pic 1 On the right is the Abuna River, on the left is the Madeira River

acapulconis, *Trigona* Strand, 1919 also relates to [Geotrigona acapulconis (Strand, 1919) *Typ. Loc.:* **Acapulco**, Mexico]* (Moure's Bee Catalogue, 2008) = Refers to **Acapulco - + - nis** (see suffix **–nis**);

A port and city on the west coast of Mexico, in the state of Guerrero.

Pic 2 A 1628 relief atlas of Acapulco bay

aequinoctialis, *Geotrigona* Ducke, 1925 **also refers to** [*Melipona* Ducke, 1925]* (Moure's Bee Catalogue, 2008) = **Latin aequinoctiālis** m, f (*neuter* aequinoctiāle); *third declension* - equinoctial (pertaining to the equinox)

aequatoriana, *Partamona* Camargo, 1980 also relates to [*Melipona* Schulz, 1904 Typ. Loc. Esmeraldas, **Ecuador**.]* (Moure's Bee Catalogue, 2008) = aequator English æquator (*plural* **æquators**) - Archaic spelling of **equator (equator** (*plural* equators) - (often "the Equator") An imaginary great circle around the Earth, equidistant from the two poles, and dividing earth's surface into the northern and southern hemisphere.) + - iana (see suffix –**iana**)

Ecuador is a small country in South America. It is along the equator. **Ecuador** officially the **Republic of Ecuador** (Spanish: *República del Ecuador* [reˈpuβlika ðel ekwaˈðor], which translates as "Republic of the Equator")

africana, *Melipona* and refers to [*Trigona africana bibundicola, Trigona africana neavei, Trigona africana rhodesica* and *Trigona africana medionigra,*]^ (Eardley, 2004) = From Latin **āfricānus** m (*feminine* **āfricāna**, *neuter* **āfricānum**); *first/second declension* - African - Of or pertaining to Africa – Latin **Āfrica** f (*genitive* **Āfricae**);

first declension - Northwestern Africa, the territory of Carthage, the African coast west of the Nile;

Africa as a continent is understood as the quarter of the globe south of the Mediterranean.

amalthea, *Trigona* (Olivier, 1789) and [*Apis amalthea* Olivier, (1789) syn. Typ. Loc. *:* Cayenne, French Guiana.]* (Moure's Bee Catalogue, 2008) = Possibly named after the Greek goddess **Amalthea**. Sp. 16

Sp. 16 *Trigona amalthea* http://commons.wikimedia.org/wiki/File:Trigona_amalthea.jpg

In Greek mythology, **Amalthea** or **Amaltheia** (Greek: Ἀμάλθεια) is the most-frequently mentioned foster-mother of Zeus. Her name in Greek ("tender goddess") is an epithet.

amapana, *Trigona (Trigona)* Schwarz, (1948). [Typ. Loc.: **Amapá,** Brazil, Ducke leg.]* (Moure's Bee Catalogue, 2008) = Refers to **Amapa**, + ana (see suffix **–ana**)

A province in Brazil (adjacent to French Guiana) The estuary of the River Oiapoque, once considered the northernmost point of Brazil, is at the extreme north of the Brazilian coast. The dominant feature of the region—90 percent of the total area—is the Amazon Rainforest.

amazonensis, *Trigona* (Ducke, 1916) [*Typ. Loc.:* Tefé and Alto Rio Purus, **Amazonas**, and Rio Acre, Acre, Brazil]* (Moure's Bee Catalogue, 2008) = Refers to **Amazonas,** South America + -ensis (see suffix **–ensis**);

Amazonas is a state of Brazil located in the northwestern corner of the country. It is the largest Brazilian and the 9th largest country subdivision in the world.

amazonica, *Melipona* Schulz, 1905 also [*Melipona marginata* Schulz, 1905: Distribution: Neotropical Region: Brazil (Acre, Amapá, **Amazonas,** Pará, Rondônia)]* (Moure's Bee Catalogue, 2008) = Amazon - pertaining to or of the area of Amazon, South America. + -ica (see suffix **-ica)** see also amazonensis.

argentina, *Geotrigona* Camargo & Moure, 1996 [Typ. Loc. *:* **Argentina,** Santiago del Estero, Las Termas, 11.X.1972,]* **Argentina**,

Officially the Argentine Republic is a federal republic located in southeastern South America.

atlantica, *Scaura* Melo, 2004 [Etym.: biome in which it has been found, the Atlantic forest of eastern Brazil.]* (Moure's Bee Catalogue, 2008) = From Latin **ātlanticus** m (*feminine* **ātlantica**, *neuter* **ātlanticum**); *first/second declension* -Atlantic;

The **Atlantic Forest** (Portuguese: *Mata Atlântica*) is a terrestrial Biome and region which extends along the Atlantic coast of Brazil from Rio Grande do Norte state in the north to the Rio Grande do Sul state in the south, and inland as far as Paraguay and the Misiones Province of Argentina.

atomaria, *Trigonisca* (Cockerell, 1917 [Typ. loc.: Pozo Azul, Costa Rica, 15.VI.1902, M. A. Carriker Jr. coll.]* (Moure's Bee Catalogue, 2008) = From Latin atomus m (*feminine* **atoma**, *neuter* **atomum**); *first/second declension* - indivisible + ria From *rio* ("river").**ria** - mouth of a river, estuary, river branch;

Of or pertaining to the estuary at Sarapiqui River, Pozo Azul Costa Rica, Central America.

australis, *Austroplebeia* (Friese, 1898) [Typ. Loc. : Australia "C.-Australien (v. Müller 1893)" "Burnett-River (Queensland, Simon 1891-92)"][#] (Rasmussen, 2008) = From *auster* ("south") + *-ālis*. Latin **austrālis** m, f (*neuter* **austrāle**); *third declension* - southern;

sp 1 *Austroplebeia australis* (Wikipedia, A. australis 2014)

Australia, officially the **Commonwealth of Australia**, is a country comprising the mainland of the Australian continent, the island of Tasmania, and numerous smaller islands

azteca, *Trigonisca* Ayala, 1999 [*Etym.*: homage to Azteca culture.]* (Moure's Bee Catalogue, 2008) = Spanish azteca – Of or pertaining to The **Aztec** people were certain ethnic groups of central Mexico.

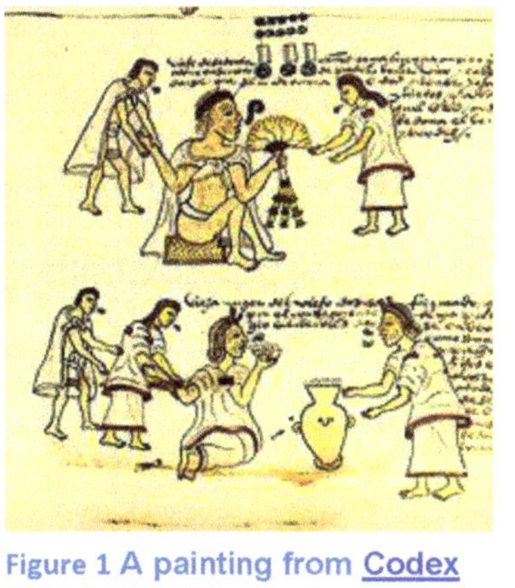

Figure 1 A painting from Codex Mendoza (Wikipedia, Aztec, 2014)

Particularly to those groups who spoke the Nahuatl and who dominated large parts of Mesoamerica from the 14th to 16th centuries. The Nahuatl words *aztecatl* [asˈtekat͡ɬ] (singular) and *aztecah* [asˈtekaʔ](plural) mean "people from Aztlan", a mythological place for the Nahuatl-speaking culture of the time, and later adopted as the word to define the Mexica people.

bahiana, *Melipona* Ducke, (1916) **[***Typ. Loc. :* Barra, Bahia, Brazil, Adolpho Lutz coll.**]*** (Moure's Bee Catalogue, 2008) = Refers to Bahia, + -ana (see Suffix –ana)

Bahia (local pronunciation: [baˈi.ɐ]) is one of the 26 states of Brazil and is located in the eastern part of the country on the Atlantic coast

baleensis, *Liotrigona* Pauly & Hora sp. nov^* (ITIS, 2014) = Refers to the **Bale Mountains** + ensis (see suffix **–ensis**) (also known as the **Urgoma Mountains**),

The Urgoma Mountains is a range of mountains in the Oromia Region of southeast Ethiopia, south of the Awash River.

barticensis, *Melipona fasciata* Cockerell, 1918. *Typ. Loc.:* Bartica, British Guiana]* (Moure's Bee Catalogue, 2008) = Bartica, a town in Guyana + ensis (see **ensis**);

Bartica developed from an Anglican missionary settlement, established in 1842. The name 'Bartica' comes from an Amerindian word meaning 'red earth', abundant in the area.

Sp. 17 *Tetragonula bengalensis Photo: Claus Rasmussen*

barrocoloradensis, *Scaptotrigona* (Schwarz, 1951) *Typ. Loc.:* **Barro Colorado** Island, Canal Zone, Panama, 23.III.1933, Lutz, F. E. leg..]* (Moure's Bee Catalogue, 2008) = Refers to **Barro Colorado** + ensis (see suffix **–ensis**);

Barro Colorado Island (BCI) is located in the artificial Gatun Lake in the middle of the Panama Canal.

bengalensis, *Tetragonula* (Cameron, 1897) [*Typ. Loc.:* India "Barrackpore (Rothney)"][#] (Rasmussen, 2008) = Refers to Bengal + **ēnsis** (see suffix **–ensis**) Sp. 17

A region in the Indian subcontinent. The name of Barrackpore, derived from the terms Barrack, is because troops have been stationed here since **1772**. The name dates back to a common village name of **Bengal.** (http://www.barrackpore.gov.in/HTM/sdobkp_History_Recent.htm)

belizeae, *Melipona* Schwarz, 1932 [*Typ. Loc.:* Belize]* (Moure's Bee Catalogue, 2008) = Concerning Belize + -ae (see suffix **–ae**)

Belize is a country on the northeastern coast of Central. Belize is bordered on the north by Mexico, south and west by Guatemala, and east by the Caribbean Sea.

betsimisaraka, *Liotrigona* Pauly, 2001 (Eardley & Urban, 2010) = Pertaining to, from or of **Betsimisaraka** (people);

A Malagasy people living along the east-central and northeastern coast of Madagascar. The **Betsimisaraka** ("the many inseparable") makeup approximately fifteen percent of the Malagasy people and are the second largest ethnic group in Madagascar

bibundicola, *Trigona africana* var. Strand, 1912^ (Eardley, 2004) = **Refers to Bibundi** + cola (see **cola**) **Bibundi** bay - **Bibundi river** a region in Cameroon, West Africa

boliviana, *Melipona* Schwarz, 1932 also [*Melipona fasciata* Schwarz, 1932 *Typ. Loc.:* Rio Colorado, **Bolivia**, September [1921-22], Wm. M. Mann. Distribution: Neotropical Region: Bolivia (*Cochabamba, La Paz*)]* (Moure's Bee Catalogue, 2008)= Refers to Bolivia, + -ana (see suffix **–ana**)

Bolivia (Spanish: [bo'liβja], Quechua: *Buliwya*, Aymara: *Wuliwya*), officially known as the **Plurinational State of Bolivia** (Spanish: *Estado Plurinacional de Bolivia* locally: [es'taðo plurinasjo'nal de βo'liβja]), Bolivia gained its Independence in 1825 after a rebellion, which was led by Simon Bolivar, who later gave his name to the country. Bolivia is one of the least developed countries in South America.

bonitensis, *Melipona* Cockerell, (1919) [*Typ. Loc.:* **Bonito**, Pernambuco, Brazil]* (Moure's Bee Catalogue, 2008) Relates to **Bonito** + -ensis(see suffix –ensis)

Bonito (*'Beautiful' in Portuguese*) is a city in northeastern Brazil, in the State of Pernambuco. It lies in the mesoregion of Agreste of Pernambuco

borealis, *Trigona amalthea* var.Friese, (1913). (Nomen nudum) [specimens from San Mateo, Costa Rica]* (Moure's Bee Catalogue, 2008) = Latin **boreālis** m, f (*neuter* **boreāle**); *third declension* - northern

borneënsis, *Trigona* Friese 1933 [*Typ. Loc.:* Indonesia "Sanggau (Borneo) 24.Juli 1932"][#] (Rasmussen, 2008) = Refers to the Island of Borneo; **borneensis** Inflection: borneens. + -ensis (see suffix –ensis)

Related terms: Borneum Usage notes: Used almost exclusively as a taxonomic epithet.The greater abundance of stingless bees in Thailand and Malaysia, including all of Borneo, is due to the abundance of resin-secreting trees (Dipterocarpaceae) and humid tropical climate, although most likely an array of factors are responsible for this concentration of species.[#]

caliensis, Parapartamona (Schwarz, 1948) [*Typ. Loc.:* **Cali**, Cauca Valley, Colombia]* (Moure's Bee Catalogue, 2008) cali + ensis (see suffix **–ensis**)

Santiago de Cali (Spanish pronunciation: [san'tjaɣo ðe 'kali]), usually known by its short name "Cali", is the capital of the Valle del Cauca department in western Colombia and is the country's third-largest city. The city was founded on 25 July 1536 by the Spanish conquistador Sebastián de Belalcázar.

cambodiensis, *Trigona* Cockerell 1926 [*Typ. Loc.:* **Cambodia** "Angkov (=Angkor) Wat, Cambodia, Jan. 11, 1926 (H.M. Smith)"][#] (Rasmussen, 2008) = Pertaining to Cambodia + **ēnsis** (see suffix **–ensis**)

Used almost exclusively as a taxonomic epithet. Cambodia (**Kampuchea**, IPA: [kɑm'pu'ciə]), officially known as the **Kingdom of Cambodia** and once known as the Khmer Empire, is a country located in the southern portion of the Indochina Peninsula in Southeast Asia.)

cameroonensis, Meliponula (Axestotrigona) (Friese) comb.n., and [Type material. *Trigona cameroonensis,* 'W Africa Togo', 'N **Kamerun** Joh: Albrechtshohe]^ (Eardley, 2004) = Refers to Cameroon + **ēnsis** (see suffix **–ensis**)

Used almost exclusively as a taxonomic epithet. Cameroon (officially the **Republic of Cameroon** (French: *République du Cameroun*) is a country in the West Central region.)

capixaba, *Melipona (Michmelia)* Moure & Camargo, 1994 [Etym: from Tupi: *kapi'xawa* = plantation ground; also the nickname for inhabitants from Espírito Santo.]* (Moure's Bee Catalogue, 2008) = "Capixaba" is a word from Tupi. It means "cabelo de milho" (Corn Hair) because when the Portuguese arrived, they were blonde, so their hair reminded the native people of the golden colour of the corn hair.

Originally, the "Capixabas" were the individuals born in Vitória only. Lately, the term spread to those born anywhere in the state of **ES**, although the precise gentilic is "espiritossantense". **Espírito Santo** (Portuguese pronunciation: [es'piritu 'sɐ̃tu]) is one of the states of southeastern Brazil, often referred to by the abbreviation "ES".

carioca, Melipona marginata Moure, 1971 [Typ.Loc. Rio de Janeiro, Brasil, 352 m, I.1951, C. A. C. Seabra leg..]* (Moure's Bee Catalogue, 2008) = Relating to the city of Rio de Janeiro, Brazil.;

Portuguese - **carioca** (*plural* cariocas) - An inhabitant of the city of Rio de Janeiro, Brazil

carvoeiroensis, *Trigona (Partamona) testacea* var. Pedro, S. R. M. & Camargo, J. M. F. (2003). (Nomen nudum), [H. F. Schwarz labeled some specimens, from **Carvoeiro**, rio Negro-rio Branco, Brazil]* (Moure's Bee Catalogue, 2008) = Refers to Carvoeiro + -ensis (See –ensis)

Carvoeiro, a village on the Rio Negro, in the municipality of Barcelos in the state of Amazonas, in Brazil;

catamarcensis, *Plebeia* (Holmberg, 1903) [*Typ. Loc.:* Catamarca, Argentina, 14.II.1899]* (Moure's Bee Catalogue, 2008) = Refers to **Catamarca**, + ensis (see suffix **–ensis**);

San Fernando del Valle de Catamarca (Spanish pronunciation: [sam fer'nando ðel 'βaʎe ðe kata'marka]) is a city in northwestern Argentina and capital of Catamarca Province, on the Río Valle River, at the feet of the Cerro Ambato. The city name is normally shortened as **Catamarca**

centralis, *Trigona* Pedro & Camargo, 2003 also refers to *Trigona (Partamona) testacea* ***centralis*** Pedro, S. R. M. & Camargo, J. M. F. (2003) Nomen nudum (Moure's Bee Catalogue, 2008) = Latin **centrālis** m, f (*neuter* centrāle); *third declension* - central, in the middle.

Refers to the Central region (see Brazil regional map)

chachapoya, *Dolichotrigona* Camargo & Pedro, 2005 [Etym: homage to the pre-inca civilization Chachapoya.]* (Moure's Bee Catalogue, 2008)

The **Chachapoyas**, also called the *Warriors of the Clouds*, was a culture of Andean people living in the cloud forests of the Amazonas of present-day Peru;

chacoana, Lestrimelitta Roig Alsina, 2010 [*Typ. Loc.:* Argentina, Chaco, J. J. Castelli, campo Don Medina, 7-I-2005, F. Vossler]* (Moure's Bee Catalogue, 2008) = Refers to the **Province of Chaco** + -ana (see suffix **–ana**)

Sp. 18

Sp. 18 *Lestrimelita chacoana* Photo: via Maria Rosa Bernasconi.

In Spanish: *Provincia del Chaco* is a province in northeastern Argentina.

chamelensis, Lestrimelitta Ayala, R. (1999). [Typ. Loc.: **Chamela** Estación Biológia, Jalisco, Mexico Etym.: refers to the type locality]* (Moure's Bee Catalogue, 2008) = Pertaining to or from the Chamela (EBCh) + ensis (see –ensis);

The Chamela Biological Station, Jalisco, Mexico, contributes a third of global knowledge on dry forests, which makes the site the greatest contributor to the understanding of these habitats.

chanchamayoensis, Trigona Schwarz, 1948 *Typ. Loc.:* San Ramon, Valle de **Chanchamayo**, Peru, 5.II.1940]* (Moure's Bee Catalogue, 2008) = Refers to **Chanchamayo,**.+ ensis (see suffix –ensis) Sp. 19.;

Sp. 19 *Trigona chanchamaoensis* Photo: Claus Rasmussen

Chanchamayo (in Hispanicized spelling) or **Chanchamayu** (Quechua *chanchay* to walk and leap about, to walk quickly and confused, *chancha chancha* to walk quickly and irregularly, *shancha* a kind of bird, *mayu* river) The Chanchamayo Valley is a province in northern Junín Region, in central Peru is situated 300 km east of Lima. San Ramon is one of 2 important cities in the region.

chapadana, Nannotrigona (Schwarz, 1938) *Typ. Loc.:* Sant'Anna da **Chapada**, near Cuyabá, Mato Grosso, Brazil]* (Moure's Bee Catalogue, 2008) = Refers to **Chapada** + na (see suffix **–ana**);

Chapada, a plateau found in the Brazilian Highlands. see also *chapadicola*

chapadicola, Partamona Pedro & Camargo, 2003 [*Typ. Loc.:* **Chapadinha**, 39 km S, Maranhão, Brasil. *Etym.:* The name refers to the region of "**chapada**", where the species inhabits.]* (Moure's Bee Catalogue, 2008) = Pertains to **chapada**, + cola (see **cola**);

Chapadinha is a Brazilian city and municipality in the state of Maranhão. It is located 235 km southeast of the state capital São Luís.

chiriquiensis, *Geotrigona* Schwarz, 1951 [*Typ. Loc.:* El **Volcan de Chiriqui**, Panama, 20.II.1936, F. E. Lutz, W. J. Gertsch leg.]* (Moure's Bee Catalogue, 2008) = Refers to Chiriquí, + -ensis (see suffix **–ensis**)

Volcán is a town in the Chiriquí province of Panama. It is located on the western coast of Panama.

Figure 2 Chiriqui National Park (Wikipedia, Chiriqui Province, 2014)

chocoana, *Oxytrigona* Gonzalez & Roubik, 2008 *Etym.:* reference to the **Chocó** region (Ecuador **Choco** rainforest).]* (Moure's Bee Catalogue, 2008) = Refers to Choco + -ana (suffix –ana)

The Choco Rainforest -- one of the world's last coastal tropical rainforests -- along the coastal lowlands of western Ecuador.

chromensis, *Liotrigona* Pauly, 2001^ (Eardley, 2004) = Refers to Morarano **Chrome** + -ensis (see suffix **–ensis**)

Morarano Chrome is a town and commune (Malagasy: '**kaominina**') in Madagascar. It belongs to the district of Amparafaravola, which is a part of the Alaotra-Mangoro Region. This place is situated in Amparafaravola, Toamasina, Madagascar.

colimana, *Melipona (Michmelia)* Ayala, 1999 [Typ. loc.: Laboratorio Natural Las Joyas, Sierra de Manantlán (al Sur de Autlán), Jalisco México, 13.VII.1988, F. A. Noguera]* (Moure's Bee Catalogue, 2008) = Pertaining to **Colima**, + ana(see suffix **-ana**)

Colima, a small state located in the Western region on Mexico's central Pacific coast. It shares borders with the states of Jalisco and Michoacán.

Figure 3 Waterfall on Gunung Ledang (Mt Ophir) Johor Malaysia (Wikipedia, Mount Ophir, 2014)

collīna, *Tetragonilla* (Smith, 1857) [Typ. loc.: Malaysia "Malacca (**Mount** Ophir)"]# (Rasmussen, 2008) = From Latin - **collīnus** m (*feminine* **collīna**, *neuter* **collīnum**); *first/second declension* - Of a hill; growing on a hill.

Mount Ophir, or more commonly known by its Malay name, **Gunung Ledang**, is a mountain situated in the Gunung Ledang National Park located in Ledang District (northwestern Johor), Malaysia. The summit is located between the border of Muar and Malacca.

cora, *Plebeia* Ayala, 1999 and *Plebeia (Plebeia) cora* Ayala, (1999). *Typ. Loc.:* Chipililla, Nayarit, México, 15.IX.1970, G. E. & R. M. Bohart. *Etym.:* the name refers to the **Cora** indigenous people.]* (Moure's Bee Catalogue, 2008)

The **Cora** (or Chora) are an indigenous ethnic group of Western Central Mexico that live in the Sierra de Nayarit and La Mesa de Nayar in the Mexican states of Jalisco and Nayarit. They call themselves *náayarite* (plural; *náayari* singular), whence the name of the present-day Mexican state of **Nayarit**

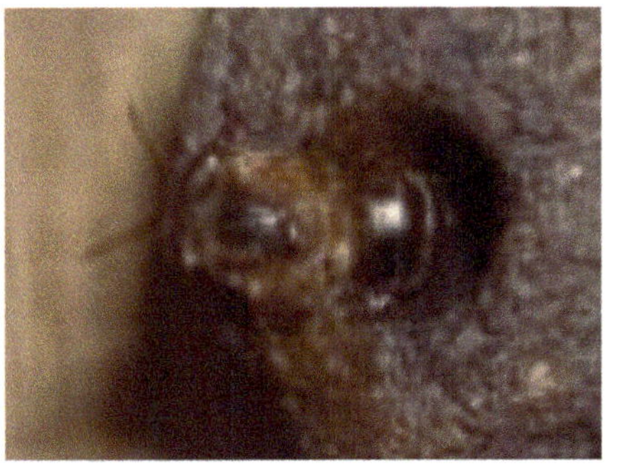

Sp. 20 *Trigona costaricensis Photo: Meliponastrigona* Costa rica

costaricensis, *Trigona* (Tetragona) Wille, A. (1962). *Typ. Loc.:* 14.5 Kms. North of Quepos, at the base of the mountains of Dota, San Jose, **Costa Rica,**]* (Moure's Bee Catalogue, 2008) = Of Costa Rica. + ensis (see suffix **–ensis**) Sp. 20

Costa Rica, meaning "rich coast" in Spanish), officially the **Republic of Costa Rica** (Spanish: *Costa Rica* or *República de Costa Rica*, pronounced: [reˈpuβlika ðe ˈkosta ˈrika]), is a country in Central, bordered by Nicaragua to the north, Panama to the southeast, the Pacific Ocean to the west, and the Caribbean Sea to the east.

coyrepu, *Melipona* Müller, (1875) Etym: From Tupi Language - alternative forms, coirepu, coeirupu - **Coeirupu** - resident in the highlands.{See Tupi glossary)

Coeirupu - resident upland - Bee Guide (Apisguia): (http://www.apisguia.com.br/?pagina=principal) (Torres, 2005)

cupira, *Partamona* (Smith, 1863) loc.: Brazil and [*Trigona cupira luteicornis* Pedro & Camargo, (2003) loc. Costa Rica also *Trigona cupira nigrior* Cockerell, (1925) loc.: Trinidad.]* (Moure's Bee Catalogue, 2008);

Cúpira is a city in the state of Miranda, Venezuela. It is the capital of Pedro Gual.

dapitanensis, *Tetragonula* (Cockerell, 1925) [Typ. Loc.: Philippines. "**Dapitan,** Mindanao (Baker, 23119)"][#] (Rasmussen, 2008) = Refers to Dapitan + **ēnsis** (see suffix **–ensis**)

The beginnings of the settlement now known as the city of **Dapitan** predates the arrival of the Spanish conquistadores on the island of Mindanao. The Subanens, a nomadic tribe of Indo-Malayan stock, were the earliest known settlers who lived along river banks or "suba", from which word they received their present tribal identity as Subanen.

darek, *Trigona (Tetragonula) minangkabau* f. Sakagami & Inoue 1985: [Nomen nudum.]# (Rasmussen, 2008) = From Minangkabau language meaning high ground. Used in naming a form.

doipaensis, *Lepidotrigona* (Schwarz, 1939) and *Trigona (Lepidotrigona) ventralis* var. ***doipaensis*** Schwarz 1939 [Typ. Loc.: Thailand "SIAM.-Doi Pa, Mai Deng, 29, 1932 (H. M. Smith)"]# = refers to Doi Pa + ensis (see suffix –ensis)

Sp. 21 *Lepidotrigona doipaensis* Photo: Azmi Bin Awang Azi

domestica, *Melipona domestica* Darwin, C. (1859) (*Nomen nudum* [Darwin referred to the nest described and illustrated by J. P. Huber (1839)])* (Moure's Bee Catalogue, 2008) = From Latin **domesticus** m (*feminine* domestica, *neuter* domesticum); *first/second declension* - Of the house; domestic; familiar; native(New Latin) Used as a species epithet.

dominicana, *Proplebeia* (Wille & Chandler, 1964) *Typ. Loc.:* Pico Diego de Ocampo near Pedro García, Palo Alto de la Cumbre region, **Dominican** Republic, amber-bearing formations.]* (Moure's Bee Catalogue, 2008) = Dominica + -ana (see suffix **–ana** (Fig 1 Dominican Amber Specimen: HYMENOPTERA Apidae: Meliponini Proplebeia dominicana (Stingless Bee) Possibly 23mya (typical of Dominican Amber)

domicilii, *Trigona mosquito* var. Rau, P. (1933) (Moure's Bee Catalogue, 2008) = From *domus* ("home, house") + *colō* ("inhabit, dwell"). – Latin **domicilium** n (*genitive* **domiciliī**); *second declension* - A habitation, dwelling, domicile, abode, home. See also domiciliorum

domiciliorum, *Trigona mosquito* var.. Schwarz, (1934) (Moure's Bee Catalogue, 2008) = see domicile + rum (From Latin **rum** meaning "**of**") Schwarz, H. F. (1934).

The social bees (Meliponidae) of Barro Colorado Island, Canal Zone. *Am. Mus. Novit.* [**nesting site**: buildings]

duidae, *Melipona* fasciata var. Schwarz, 1932). *Typ. Loc.:* Middle Camp on the way from Esmeraldo [=La Esmeralda] to Mt. Duida, Venezuela, 4-7.XI.1928]* (Moure's Bee

Catalogue, 2008) = Pertaining to Mt. Duida, Venezuela with the Latin **ae** terminate (see suffix -ae);

Cerro Duida, also known as **Cerro Yennamadi**, is a very large tepui(plateau) in Amazonas state, Venezuela. At its foot lies the small settlement of La Esmeralda, from which the mountain can be climbed

dulitae, *Trigona haematoptera* variety Schwarz,1937 [&] (Schwarz H. F., 1939) = Pertaining to Mt. Dulit in Borneo with the Latin **ae** terminate (see suffix -ae);

Mount Dulit peaks at 1,311 metres (4,301 ft) above sea level and stands at the head of the Baram River in northern Sarawak, Malaysia.

eburnensis, *Trigona (Axestotrigona)* Darchen 1970, Syn. nov. *Meliponula* ***eburnensis*** Nomen nudum Type material: Côte d'Ivoire, Lamto] ^# (Eardley & Urban, 2010) = From Latin **eburnea** – (ivory) + ensis (see suffix –ensis)

Ivory Coast, officially the **Republic of Côte d'Ivoire**, is a country in West Africa.

essingtoni, *Austroplebeia* (Cockerell, 1905) Typ. Loc.: Australia Port **Essington**, N. Australia]# (Rasmussen, 2008)

Port Essington is an inlet and historic site located on the Cobourg Peninsula in the Garig Gunak Barlu National Park in Australia's Northern. It was the early attempt at British settlement but now exists only as a remote ruins series.

essequiboensis, *Tetragona* (Schwarz, 1940) *Typ. Loc.:* Upper Essequibo River, Guyana, 23.XII.1937]* (Moure's Bee Catalogue, 2008) = Refers to Essequibo River + ensis (see suffix –ensis)

Essequibo, a Dutch colony on the **Essequibo** River in the Guiana region on the north coast of South America from 1616 to 1814.;

gabonensis, *Trigona (Axestotrigona) erythra* var. Darchen 1969 [nomen nudum]^ (Eardley, 2004) = Refers to.Gabon + ensis (see suffix **–ensis**)

Gabon, officially the **Gabonese Republic**, is a sovereign state on the west coast of Central

gambiana, *Meliplebeia* Moure, 1961 and *Meliponula (Meliplebeia) gambiana* Moure comb. n. (Eardley & Urban, 2010) = Refers to **Gambia** + ana (see suffix **–ana**)

The Republic of the **Gambia**, also commonly known as the **Gambia**, is a country in West Africa. It is surrounded by Senegal, apart from a short strip of Atlantic coastline at its western end. It is the smallest country in mainland Africa. The country is situated on either side of the Gambia River, the nation's namesake.

gambiensis, *Axestotrigona ferruginea* var. Moure, 1961^ (Eardley, 2004) = Refers to **Gambia** + -ensis (see suffix **–ensis**) see also *gambiana*

grapiuna, *Plebeia* Melo & Costa, 2009 [Typ. loc.: Brasil, Bahia, Ilhéus, Campus da UESC, 17.i.2007, Melo & Costa, em ninho. Etym.:The name 'grapiuna' is applied to the cocoa-producing region of southern Bahia, as well as a nickname for its inhabitants; from the Tupi-Guarani guirá, bird, plus piuna, dark skin, used herein allusion to the dark integument of the worker bees.]* (Moure's Bee Catalogue, 2008)

guatemalensis, *Paratrigona* (Schwarz, 1938) [*Typ. Loc.:* Chiquimulilla, Santa Rosa, **Guatemala**,]* (Moure's Bee Catalogue, 2008) = Refers to Guatemala + -ensis (see suffix **–ensis**)

Officially the **Republic of Guatemala,** a country in Central America.

guerreroensis, *Melipona fasciata* var. Schwarz, (1936) [Typ. Loc.: Chilpancingo, Guerrero, Mexico, H. H. Smith]* (Moure's Bee Catalogue, 2008) = Refers to Guerrero + -ensis (see suffix **–ensis**)

Officially **Free and Sovereign State of Guerrero** is located in Southwestern Mexico. The state was named after Vicente Guerrero

guianae, *Trigona* Cockerell, 1910 [Typ. Loc.: Guiana] = Latin - Guiana f (*genitive* **Guianae**) (Moure's Bee Catalogue, 2008); *first declension* – or (the) Guianas (plural only)

The Guianas, sometimes called by the Spanish loan-word **Guyanas** (Las Guayanas), refers to a region in north-eastern South America which includes the following three territories: **1) French**, an overseas department of France **2) Guyana, known as British** from 1831 until 1966, after the colonies of Berbice, Essequibo, and Demerara, taken from the Netherlands in 1814, were merged into a single colony **3) Suriname**, until 1814 a part of Dutch Guiana, together with Berbice, Essequibo, and Demerara

gurupina, *Melipona* Müller, H. (1875) (Nomen nudum)* (Moure's Bee Catalogue, 2008) = gurupu or guarupu is a folk name for (*Melipona (Eomelipona) bicolor schencki* Gribodo, 1893) (Krug, 2010) + -ina(see suffix –ina)

guyanensis, *Lestrimelitta* Roubik, D. W. (1980) [Typ. Loc.: Kourou, 17 km southwest, French Guiana, D. Roubik leg., robbing nest of Apis mellifera.]* (Moure's Bee Catalogue, 2008) = pertaining to or of French Guiana, South America + -ensis (see suffix **–ensis**)

A country in South America. Official name: Cooperative Republic of Guyana. see also guianae

sp 2 *Lepidotrigona hoozana*
Photo: I Hsing Sung

***hoozana** Lepidotrigona* (Strand, 1913), comb. nov. also [*Trigona ventralis hoozana* Strand 1913b: Type locality: Taiwan Hoozan]# (Rasmussen, 2008) = Taiwanese racial name was given Strand 1913 – Hoozan = Fengshan, Kaohsiung County Taiwan.

Fengshan city (a.k.a. **Hōng-soaⁿ** in Taiwan Amoy Hokkien dialect), in Koahsiung County.

***huaoranii** Oxytrigona* Gonzalez & Roubik, 2008 [Typ. loc.: Ecuador, Napo, Yasuni Nat. Park Est. La Católica, 14-26.IV.1998, D. Roubik leg. *Etym.:* homage to the **Huaorani** people, from Napo region, Ecuador]* (Moure's Bee Catalogue, 2008)

The **Huaorani**, **Waorani** or **Waodani**, also known as the **Waos**, are native Amerindians from the Amazonian Region of Ecuador (Napo, Orellana and Pastaza)

***huilensis**, Lestrimelitta* Gonzalez & Griswold, 2012 [Typ. loc.: Huila, Rivera, Cementerio, 29.mar.1994, 745 m, G. Nates. Leg. *Etym.:* The specific epithet refers to Huila, the department in Colombia where the species is currently known to occur.]* (Moure's Bee Catalogue, 2008) + ensis (see suffix –ensis)

Huila (Spanish pronunciation: [ˈwila]) is one of the departments of Colombia. It is located in the southwest of the country.

***hypogea** Trigona* Silvestri, 1902 [nesting site: subterranean]* (Moure's Bee Catalogue, 2008) and *Trigona (Trigona)* ***hypogea** robustior* Schwarz, 1940 = From English **hypogeum** (plural hypogeums or hypogea) - An underground room or cavern (also used figuratively). Sp. 22

Sp. 22 *Trigona hypogea* Photo: via Maria Rosa Bernasconi

***insularis**, Melipona (Melikerria)* Roubik & Camargo, 2012 [Typ. loc.: Panama: Veraguas Prov. Coiba Island, Playa Hermosa, 1.6 km W, 17 December 2007, D. Roubik. Etym.: from Latin, *insularis*, isolated, island-dwelling.]* (Moure's Bee Catalogue, 2008)

Coiba is the largest island in Central, with an area of 503 square kilometres, off the Pacific coast of the Panamanian province of Veraguas

***irianensis**, Trigona (Pacificotrigona) okazawai* var. Tadauchi *et al.* 1998: Nomen nudum# (Rasmussen, 2008) = Irian - pertaining to or of the Island of Irian, Indonesia. + ensis (see suffix ensis)

The name is unavailable for taxonomic purposes, as no description or definition of the taxon was provided (ICZN, art. 15.1).

isthmina, Oxytrigona Gonzalez & Roubik, 2008 Panamá, Darién, El real-Pijivasal, 16.II.1989, D. Roubik *Etym.:* reference to the Panamá Isthmus.]* (Moure's Bee Catalogue, 2008) = isthmi From Latin **isthmus** (*plural* isthmuses *or* isthmi) - A narrow strip of land, bordered on both sides by water, and connecting two larger landmasses. + ana(see suffix –ana);

javanica, *Lepidotrigona* (Gribodo, 1891) [Typ. loc.: Indonesia "Giava"][#] (Rasmussen, 2008) = From or pertaining to or of the Island of Java, Indonesia. + -ica (see suffix –ica)

Formed mostly due to volcanic, Java is the 13th largest island globally and the fifth-largest island in Indonesia.

sp 3 *Lepidotrigona javanica* Photo: Juanda Iyong

jatiformis, *Plebeia* (Cockerell, 1912 and syn *Trigona jatiformis* Cockerell, (1912)* (Moure's Bee Catalogue, 2008) = From jati (see jaty) + formis (from Latin forma "appearance")

Sp. 23 *Tetragonisca angustula* Photo : Luiz Carlos Da Cunha

jaty, *Trigona* Smith, F. (1863). [Typ. loc.: Brazil]* (Moure's Bee Catalogue, 2008) = jaty or jatai is a folk name for *Tetragonisca angustula* (Latreille, 1811) in South America. (Krug, 2010); Sp. 23

From Tupi Language **Jati** (nickname Jatahy) - yellow sweet honey bees.

jombenensis, *Trigona africana* Cockerell and also refers to [*Trigona beccarii jombenensis* Cockerell, 1917Syn.n. Type material: Jombene Range, East Africa (Chanler- Hohnel Expedition)][^&] (Cockerell, New Social Bees., 1917) = From or of Jombene + ensis (see suffix –ensis)

sp 4 *Trigona jaty* (Wikipedia, Abelha-jatai 2009)

jujuyensis, *Scaptotrigona* (Schrottky, 1911) *Trigona jujuyensis* Schrottky, (1911). *Typ. Loc.:* San Pedro, Provincia de **Jujuy**, Argentina, 24-IV-1981]* (Moure's Bee Catalogue, 2008) = Refers to Jujuy + ensis (see suffix ensis) Sp. 24

Jujuy, a province of Argentina, is located in the extreme northwest of the country, at the borders of Chile and Bolivia.

kaba, *Geotrigona* Gonzalez & Sepúlveda, 2007 [Typ. loc.: Colombia, Antioquia, Porce, Tenche. Pescado, 24.VIII.1997, Allan Smith leg. Etym.: Kaba, from U'wa mythology, the spirit which dominates the underground world.]* (Moure's Bee Catalogue, 2008)

Sp. 24 Scaptotrigona jujuyensis Photo: Maria Rosa Bernasconi

kaieteurensis, *Tetragona* (Schwarz, 1938) and [*Trigona (Tetragona) kaieteurensis* Schwarz, (1938).Typ. Loc.: **Kaieteur**, British Guiana, 27.V.1929 Miss Alice Mackie]* (Moure's Bee Catalogue, 2008) = Refers to Kaieteur + ensis (see suffix **-ensis**).

Kaieteur Falls is a high-volume waterfall on the Potaro River in central Essequibo Territory, Guyana. It is located in **Kaieteur** National Park.

kangarumensis, *Melipona lateralis* Var. Cockerell, (1920) [Typ. Loc.: Kangaruma, Guyana, 18.VIII.1911, Lutz, forest near this place]* (Moure's Bee Catalogue, 2008) = Refers to **Kangaruma**, + ensis (see suffix **–ensis**)

Kangaruma is a village settlement in Guyana, South America

katagensis, *Trigona denoiti* var. Cockerell, 1934 and *Plebeina denoiti katagensis* (Cockerell) Type Material: 'Katanga Dilolo 27.vii.1931 J Ogilvie'; 'Katanga Lundu vi.1924 Ch Seydel' ; 'Kapiri ix.1912 Miss Agric.'; 'Katanga, Tenke, 30.vii–9.viii.1931, J Olgilvie']^ (Eardley, 2004) = Refers to Katanga + ensis (see suffix **–ensis**)

Katanga is one of the provinces of the Democratic Republic of the Congo. Between 1971 and 1997 (during the rule of Mobutu Sese Seko when Congo was known as Zaire), its official name was **Shaba Province.** The province forms the Congolese border with Angola and Zambia. Although Katanga province and Tanzania do not share a land border, the province also borders Tanzania, but the border is within Lake Tanganyika.

keyensis, *Platytrigona* (Friese, 1901) [Typ. loc.: Indonesia "Key-Eilanden (Amboina, Nederland India) durch(via) H. Kühn"]# (Rasmussen, 2008) = Refers to Key-Eilanden(Ambon Island) + ensis (see suffix **-ensis**).

Key-Eilanden (Amboina, Nederland India), Ambon, Indonesia The original name *Dutch Indies* (Dutch: *Nederlandsch-Indië*) was translated by the English as the *Dutch East Indies*. Under the Dutch Empire, Ambon city was the seat of the Dutch resident and military commander of the Moluccas.

khasiana, *Melipona* **Pugh** 1947 [Nom nud.]# (Rasmussen, 2008) = **Khasi** (plural **Khasis** or **Khasi**) - A member of an Indian tribe, the majority of whom live in Meghalaya, India + ana (see suffix **–ana**)

The name is unavailable for taxonomic purposes, as no description or definition of the taxon was provided (ICZN, art. 13.1.1) (common name (nagp hamang)#

kusutkana, *Trigona* Dover 1929 [Typ. Loc.: Malaysia "Kedah Peak."]# (Rasmussen, 2008) kusutkana refers to the Riau Malay phrase "kusut kana", meaning "tussled or ruffled because of" implying hair being tussled or ruffled after an attack of a swarm of these bees.

The language of the Riau Islands is known as Riau Malay. The Riau Islands have been considered the birthplace of the modern Malay language. The Banjar language carries the same meaning for this phrase. Modern Malay pronounces the phrase as "kusut kerana".

komiensis, *Trigona nebulata* subsp. Cockerell, 1934a: [Typ.loc.: 'Sankuru, Komi, 30.iii. 1930]^ (Eardley, 2004) - Named after the region this species occurs, Komi in the Democratic Republic of Congo, Africa. + ensis (see suffix –ensis)

kwyrakai, *Geotrigona* Camargo & Moure, 1996 [Typ. loc: Brazil, Pará, Jacareacanga, X.1959, M. Alvarenga leg.; Etym.: homage to the Kayapó Indian Kwirà-kà [*sic* = Kwyrà-kà].]* (Moure's Bee Catalogue, 2008) "

The noted bee expert of the Gorotire-Kayapó, Kwyrá-ká, for example, was able to separate for us quickly and accurately numerous closely related *Trigona* species but was unable to verbalize the reasons for his decisions. This is because his knowledge of nature comes from silently observing rather than verbally analysing. Observations are registered in a gestalt manner, along with a myriad of other information regarding niche, habitat, ecological zone, geographic coordinates and associated elements in the same ecosystem."

lamingtonia, *Platytrigona* (Cockerell, 1929)[Typ. loc.: Papua New Guinea "Mt. Lamington, May 1927 (C.T.McNamara)"]# (Rasmussen, 2008) = pertaining to or of Mt. Lamington. + ia (see suffix **–ia**)

Mount Lamington is an andesitic stratovolcano in the Oro Province of Papua. The forested peak of the volcano had not been recognised as such until its devastating eruption in 1951. The mountain was named after Charles Cochrane-Baillie, 2nd Baron Lamington, who was Governor of Queensland.

limae, Scaptotrigona (Brèthes, 1920) [*Typ. Loc.:* **Lima**, Peru.]* (Moure's Bee Catalogue, 2008)

Lima Province is located on the central coast of Peru and is the only province in the country not belonging to any of the twenty-five regions. Its capital is Lima, which is also the nation's capital.

madecassa *Liotrigona* (Saussure, 1890) - The specific epithet*madecassa is d*erived from the French word *Madecasse* for, indigenous' or 'inhabiting Madagascar'.*+ (Smith & Walker, 2011)

Madagascar, officially the **Republic of Madagascar** (Malagasy: *Repoblikan'i Madagasikara* [republiˈkʲan madagasˈkʲarə̥]; French: *République de Madagascar*) and previously known as the **Malagasy Republic**, is an island country in the Indian Ocean, off the southeastern coast of Africa. The nation comprises the island of Madagascar (the fourth-largest island in the world) and numerous smaller peripheral islands.

mahafalya, *Liotrgona* (Moure & Camargo) (Eardley & Urban, 2010) – Refers to a dialect Madagascar's Malagasy language.

The **Mahafaly** (an ethnic group of Madagascar that inhabits the plains of the Betioky-Ampamihy area)

manantlensis, *Plebeia (Plebeia)* Ayala, R. (1999) [Typ.loc.: Laboratorio Natural de las Joyas, Sierra de Manantlán (SE from Autlán), Jalisco, México, F. A. Noguera & A. Rodríquez. *ETYM.:* reference to the Sierra de Manantlán]* (Moure's Bee Catalogue, 2008) = Manantlan + -ensis (see suffix **–ensis**)

Sierra de Manantlán reserve is located in the transition of the Nearctic and Neotropical realms and encompasses parts of the Sierra Madre del Sur, with a wide range of altitudes, climates and soils. The effects of tectonic and volcanic activities and erosion are notable within the reserve. Different types of forests are present in reserve, including mesophytic, cloud, and dry deciduous and semi-deciduous tropical forests.

manauara, *Celetrigona* Camargo & Pedro,. (2009). [Typ. loc.: Manaus, Boulevard, AM, Brasil, 1239, ninho 1, 14.IV.1986, M. V. B. Garcia.; Etym.: Manauara is the common denomination to the people born in Manaus, Amazonas, Brazil, the type locality of the species here described.]* (Moure's Bee Catalogue, 2008) = Refers to **Manaus**, + ara (**āra** f (*genitive* **ārae**); *first declension* - altar, sanctuary, refuge);

Manaus is a **city** in **Brazil**, the **capital** of the state of **Amazonas**.

manaosensis, *Melipona interrupta* var. [*Typ. Loc.:* "Mañaos" [= Manaus], Amazonas, Brazil, 11.IX.1924, J. Bequaert]* (Moure's Bee Catalogue, 2008) = Refers to Manaus + ensis (see suffix –**ensis**) See also ***manauara***

mandacaia, *Melipona (Melipona)* Smith, 1863 (Moure's Bee Catalogue, 2008) = pertaining to or of Mandacaia, in Pernambuco province, East Brazil.; From Tupi Language - **Mandaçaia** - bee fly, very gentle.

maracaia, *Lestrimelitta* Marchi, P. & Melo, G. A. R. (2006). Typ. Loc.: Ilha de Maracá, Roraima, Brazil; Etym.: reference to the type locality, "Ilha de Maracá".]* (Moure's Bee Catalogue, 2008)

Maracá Ecological Station is an ecological station some 5 km offshore of the Brazilian town of Amapá, in the state of the same name, Amapá. The ecological station is entirely located on the two adjacent islands *Ilha de Maracá do Norte*, and *Ilha de Maracá do Sul*.

marombo, *Liotrigona* n.sp.: [Etym.Refers to Marombo, or Loango (Congo, West African Coast) people passe where they dwell. Ref: The African Pygmies By A. Werner (September 1890)]~* (WERNER., 1890)

Loango city on the Atlantic coast, Founded in the 12th century as one of a cluster of Equatorial African kingdoms, Loango was in full engagement with Europeans and global trade by the 16th century.

mattogrossensis, *Melipona subterranea* var. Ducke, A. (1925). Typ. Loc.: northern Mato Grosso, upper Xingu region and tributaries of Madeira river, Brazil.]* (Moure's Bee Catalogue, 2008) = Refers to **Mato Grosso**, + **ēnsis** (see suffix **–ensis**)

sp 5 *Trigonisca maya* Photo: R. Ayala

Mato Grosso is one of the states of Brazil, the third-largest by area, located in the western part of the country

maya, *Trigonisca* Ayala, 1999 [*Etym.:* homage to Maya culture.]* (Moure's Bee Catalogue, 2008)

The **Maya** is a Mesoamerican civilization, noted for being the only known fully developed written language of the pre-Columbian Americas

sp 6 *Tetragona mayarum* Photo: John S Ascher

mayarum, *Tetragona* (Cockerell, 1912) and *Trigona ziegleri mayarum* Cockerell, (1912). Typ. loc.: Quirigua, Guatemala, W. P. Cockerell. (Moure's Bee Catalogue, 2008) = Refers to **maya** (see ***maya***) + rum (From Latin **rum** meaning "**of**")

meridionalis, *Plebeia* (Ducke, 1916) and [*Trigonisca meridionalis* Albuquerque & Camargo, 2007 *Etym.:* from the Latin, *meridionale*, inhabiting the southern region.]* (Moure's Bee Catalogue, 2008)

mexica, *Plebeia* Ayala, 1999 *Typ. Loc.:* Cuernavaca, 9.6 km E, Morellos, Mexico, 12.IX.1973, *Etym.:* homage to the Mexica people.]* (Moure's Bee Catalogue, 2008)

Spanish **Mexica** (*plural* **Mexicas**) Of or pertaining to **Mexica** - An indigenous people of central Mexico.

mexicana, *Scaptotrigona* (Guérin, 1844) [*Typ. Loc.: Mexico.*]* (Moure's Bee Catalogue, 2008) = From Latin **mexicānus** m (*feminine* **mexicāna**, *neuter* mexicānum); *first/second declension* - Mexican; from or pertaining to **Mexico**.

Mexico, officially the **United Mexican States,** also referred to as the **United States of Mexico** (Spanish: *Estados Unidos de México*), is a federal constitutional republic in North America. It is bordered on the north by the United States; on the south and west by the Pacific Ocean; on the southeast by Guatemala, Belize, and the Caribbean Sea; and on the east by the Gulf of Mexico.

minangkabau, *Tetragonula* (Sakagami & Inoue, 1985) - refer an ethnic people or to the highlands of Sumatra Indonesia. Also relates to [*Trigona (Tetragonula) minangkabau* Sakagami & Inoue 1985: Typ. Loc.: Indonesia "Lubuk Mintrun nr. Padang, Sumatera Barat, Indonesia,]# (Rasmussen, 2008) Sp. 25

The name Minangkabau is thought to be a conjunction of two words, *minang* ("victorious") and *kabau* ("buffalo").

Sp. 25 *Tetragonula minangkabau* Photo: Ahmad Zaini

mixteca, *Trigonisca* Ayala, 1999 Etym.: homage to Mixteca culture.]* (Moure's Bee Catalogue, 2008)= Pertaining to **La Mixteca** is a region in Western Oaxaca and neighbouring portions of Puebla, Guerrero in south-central Mexico which refers to the home of the Mixtec.

mombuca, *Geotrigona* (Smith, 1863) and [*Trigona mombuca* Smith, F. (1863) nesting site: obligatory subterranean]* (Moure's Bee Catalogue, 2008) = **Refers to Mombuca**, a municipality in the state of São Paulo in Brazil.;

From Tupi Language **Mombuca** - a hole in the ground, stick – suggesting subterranean nesting type species.

sp 7 *Trigona muzoensis* (Wikipedia, T. muzoensis 2014)

mondury, *Melipona (Michmelia)* Smith, 1863 [Typ. loc.: Brazil (probably from southeastern Brazil: Rio de Janeiro, eastern Minas Gerais)]* (Moure's Bee Catalogue, 2008) = From Tupi Language Mandari or **Manduri** - little nest.

Melipona mondury occurs in the Atlantic Rain Forest ranging from Santa Catarina to the Bahia States.;

muzoensis, *Trigona* Schwarz, 1948 *Trigona (Trigona) pallida* Schwarz, 1948 [*Typ. Loc.:* **Muzo**, Boyaca, Colombia, 900 m., 1936]* (Moure's Bee Catalogue, 2008) = Refers to **Muzo**. + ensis (see suffix **–ensis**).;

Muzo is a town and municipality in Boyacá, Colombia, part of the Western subregion.

nhambiquara Partamona Pedro & Camargo, 2003 [*Etym.:* homage to the Nhambiquaras Indians, from northern Mato Grosso, Brazil.]* (Moure's Bee Catalogue, 2008)

Nhambiquara is a populated place in Estado de Mato Grosso (Mato Grosso), Brazil (South America), located at an elevation of 405 meters above sea level

nitkib, *Lestrimelitta* Ayala, 1999 [Etym.: from *niitkib*, the Mayan folk name of the species.]* (Moure's Bee Catalogue, 2008) = From Mayan "nit-kib", meaning nose of wax.

L. nitkib is the only cleptobiotic (robber that obtains food and nesting materials through exclusively raiding nests of another social bee) meliponine of the Yucatan Peninsula.

nordestina, *Camargoia* Camargo, 1996 [Typ. loc.: Uruçui-Una, Bom Jesus, Piauí, Brasil, 4-10.III.1984, Camargo & Mazucato leg]* (Moure's Bee Catalogue, 2008) = From **nordestino**, a Portuguese translation of the word northeastern ("of, related to, located in, or from the northeast").

oaxacana, *Cephalotrigona oaxacana* Ayala, R. (1999 [Typ. loc.: Mexico, Oaxaca, 9.6 Km E Totolapan, 21.IX.1983, W. J. Hanson & B. A. Haws. *Etym.:* Reference to the state of Oaxaca, Mexico.]* (Moure's Bee Catalogue, 2008)

Oaxaca (English /wəˈhɑːkə/ *wə-HAH-kə*, Spanish: [waˈxaka] from Nahuatl: *Huaxyacac* [waːsʃakak]), officially **Free and Sovereign State of Oaxaca** (Spanish: *Estado Libre y Soberano de Oaxaca*), is one of the 31 states which, along with the Federal District, make up the 32 federative entities of Mexico.

occidentalis *Ptilotrigona* (Schulz, 1904) [Typ. Loc.: Nordwest-Ecuador, bei Cachavi (probably San Javier de Cachabí), 500 Fuss, XI-XII, 1896, W.F.H. Rosemberg]* (Moure's Bee Catalogue, 2008) = From Latin *occidēns* ("west") + *-ālis*. - **occidentālis** m, f (*neuter* occidentāle); *third declension* - western, westerly

ogouensis, *Meliponula (Meliplebeia)* Vachal, and *Melipona (Trigona) ogouensis* Vachal, 1903 (Eardley & Urban, 2010) = Refers to Ogou + **ēnsis** (see suffix **–ensis**)

The prefecture of Ogou, located in the Plateaux Region of Togo. (officially the **Togolese Republic** (French: *République Togolaise*), is a country in West Africa)

opita, *Lestrimelitta* Gonzalez, V. H. & Griswold, T. L. (2012). [*Typ. Loc.:* COLOMBIA: Tol. (Tolima), Armero, malaise trap, 26-20-I-1977, Peyton & Suarez. Etym.: specific epithet is a self-referential nickname for inhabitants of the department of Tolima, the area where this species is currently known.]* (Moure's Bee Catalogue, 2008)

orizabaensis, *Partamona* (Strand, 1919) [*Typ. Loc.:* Orizaba, Mexico.]* (Moure's Bee Catalogue, 2008) = Refers to **Orizaba** + ensis (see suffix – **ensis**) **Orizaba** a city and municipality in the Mexican state of Veracruz

Sp. 26 *Lepidotrigona palavanica* Photo: Zahari Deraman

palavanica, *Lepidotrigona* (Cockerell, 1915) [Type locality: Philippines "P. Princesa, Palawan (Baker coll 3839)"][#] (Rasmussen, 2008) = Refers to Palawan + -ica (see suffix –ica) Sp. 26

Palawan, officially the Province of Palawan, is an island province of the Philippines.

pampana, *Trigona* Strand, 1910 [*Typ. Loc.:* Peru, Pampa, Camona, V.1907]* (Moure's Bee Catalogue, 2008) = Refers to **Pampa** + ana (see suffix **–ana**)

Pampas is a city in Peru. It is the capital of the Tayacaja Province, and it was established on June 21, 1825

panamaensis, *Scaptotrigona* (Cockerell, 1913) *Typ. Loc.:* Las Sabanas, **Panama**]* (Moure's Bee Catalogue, 2008) = Refers to **Panama** + ensis (see suffix –ensis).

Probably from an unknown Guarani word, traditionally said to mean "place of many fish." Originally the name of the settlement founded 1519), officially **Republic of Panama** (Spanish: *República de Panamá* [reˈpuβlika ðe panaˈma]), is the southernmost country of Central.

panamica *Melipona fasciata* var. Cockerell, T. D. A. (1919). [*Typ. Loc.:* Porto Bello, **Panama**.]* (Moure's Bee Catalogue, 2008) = Relating to Panama + -ica (see suffix – ica).

Panamica also is a low stingless nettle of Central and South America having velvety brownish-green toothed leaves and clusters of small green flowers (Exact synonyms: **Friendship Plant**, **Panamiga**, **Pilea involucrata**); see also *panamensis*

paraense, *Trigona (Scaptotrigona)* Kerr et al. (1963) [(Nomen nudum) From Rio Formoso, **Pará**, Brazil)]* (Moure's Bee Catalogue, 2008) = see also ***paraensis***

paraensis, *Melipona (Michmelia)* Ducke, 1916 [*Typ. Loc.:* **Pará**, Brazil]* (Moure's Bee Catalogue, 2008) = Latin **paraensis** m, f (*neuter* **paraense**); *third declension* - Pará (attributive); Refers to **Pará,** + ensis (see suffix –**ensis);**

Para is a State in northern Brazil that has Belém as its capital.

Sp. 27 *Trigona apicalis peninsularis* subsp. Photo: by author

paraguayaca, *Melipona* Strand, E. (1910).[*Typ. Loc.:* Asuncion, **Paraguay**]* (Moure's Bee Catalogue, 2008) Refers to Paraguay + -aca (see suffix –**aca**)

Paraguay lies on both banks of the Paraguay River, which runs through the centre of the country from north to south. Due to its central location in South America, it is sometimes referred to as *Corazón de América* ("Heart of America").officially, the **Republic of Paraguay**

paulomajor, *Melipona santhilarii* var. Lepeletier, (1836) [*Typ. Loc.:* "Du Nord de La Capte. de St. Paul" [São Paulo, Brazil].]* (Moure's Bee Catalogue, 2008) = **paulo** from Latin **paulus** m (*feminine* **paula**, *neuter* **paulum**); ***first/second declension*** – "**little, small**" + **major** from Latin **māior** (*comparative of* **magnus**) – "greater, larger" – (Small but great)

penangensis, *Tetragonula* (Cockerell, 1919) [Typ. loc. : Malaysia "Penang Island (Baker 9075)"]# (Rasmussen, 2008) = Refers to Penang + **ēnsis** (see suffix –**ensis**)

A State in western Malaysia which has George Town as its capital., Island in Penang state. Founded by Francis Light and named after the Pinang Palm.

peninsularis, *Tetrigona* (Cockerell, 1927) and *Trigona apicalis peninsularis* subsp. n. Cockerell 1927: Typ. loc.: Malaysia "Perak, F.M.S., Batang Padang, Jor Camp, June 4, 1923 (Pendlebury)"]# (Rasmussen, 2008) = peninsula (Coined by Livy (59 BCE – 17 CE): *paene* ("nearly", "almost") + *īnsula* ("island")) + -aris (see suffix –**aris**) Sp. 27

Pertaining to or of a peninsular, specifically in this case the Malayan Peninsula.

sp 8 *Ptilotrigona pereneae* Photo: John S. Ascher

pereneae, *Ptilotrigona* (Schwarz, 1943) [*Typ. Loc.:* El Campamiento, Colony of the **Perene**, Peru.]* (Moure's Bee Catalogue, 2008) = Refers to **Perene** + ae (see suffix **-ae**);

The Perené River (Spanish: Río Perené) is a Peruvian river on the eastern slopes of the South American Andes.; Perené District is one of six districts of the province Chanchamayo in Peru.

peruviana, *Melipona (Melipona)* Friese, 1900 [Typ. Loc.: Callanga, Peru.]* (Moure's Bee Catalogue, 2008) = From Latin **peruvianus** m (*feminine* **peruviana**, *neuter* **peruvianum**); *first/second declension* - Peruvian (Of, from, or pertaining to Peru or the Peruvian people.) + -ana (see suffix –ana)

The Peruvian population, estimated at 30.4 million, is multiethnic, including Amerindians, Europeans, Africans, and Asians.

peruvicola, *Plebeia* Moure, 1994 [*Typ. Loc.:* Puerto Maldonado, Madre de Dios, Peru, 31.IX.1962, *Etym.:* the name refers to the country of origin]* (Moure's Bee Catalogue, 2008) = Latin **peru** (*see* also **peruviana**) + cola (see **cola**);

The word *Peru* is derived from *Birú*, the name of a local ruler who lived near the Bay of San Miguel, Panama, in the early 16th century. When Spanish explorers visited his possessions in 1522, they were the southernmost part of the New World yet known to Europeans. Thus, when Francisco Pizarro explored the regions farther south, they came to be designated Birú or Peru.

petropolis, *Trigona (Paratrigona)* Schwarz, H. F. (1948). [Typ.loc.:Petrópolis, Rio de Janeiro, Brasil, 24-27.X. 1919, Cornell Univ. Expedition]* (Moure's Bee Catalogue, 2008) = petro (Latin – "stone") + polis (Latin polus –"pole")

Petrópolis also known as **The Imperial City** of Brazil, is a city in the state of Rio de Janeiro.

picadensis, *Melipona* Strand, E. (1910) [Typ. Loc.: Picada, Salto Yguazu, Paraguay]* (Moure's Bee Catalogue, 2008) = Refers to Picada + -ensis (see suffix –ensis)

Picada, in Salto Yguazu or Iguazu Falls region is a tri-border area with Brazil, Argentina and Paraguay

piedemontana, *Lestrimelitta* Gonzalez, et al. (2010). [Typ. loc. Colombia, Caquetá, Florencia, corregimiento El Caraño, Vereda Villaraz, vía Florencia a Suaza; Etym.: reference to the geographic range, in the "piedemonte" of Colombian Andes.]* (Moure's Bee Catalogue, 2008) = Pertaing to Piedemonte + -ana (see suffix **–ana**)

sp 9 *Trigonisca pipioli* Photo: R. Ayala

Piedmonte(foothills) - Gently sloping land at the foot of a mountain range and consists of materials from erosion.

pippioli, *Trigonisca* Ayala, 1999 [Etym.: "pipioli" is the folk name in the language Nahuatl from Azteca culture.]* (Moure's Bee Catalogue, 2008)

Nahuatl is a language or group of languages of the Uto-Aztecan language family. Nahua people, most of whom live in Central Mexico; some who live in El Salvador are known as the **Pipil** or Nawat people. All Nahuan languages are indigenous to Mesoamerica. Nahuatl has been spoken in Central since at least the 7th century AD

prisca, *Cretotrigona* (Grimandi, 1988). [Alt. comb.: *Trigona prisca* ; Type specimen: AMNH C88720, an exoskeleton.; Typ. loc.: Delaware River, Kinkora, which is in a Maastrichtian terrestrial amber in New Jersey]?* (Cebaehren, 2012) = Latin prisca – meaning Empress.

Jersey Fossil stingless bee, found in New Jersey amber, late, about 80 mya.

pseudojatiformis, *Trigona* Wille. & Michener (1973) (Moure's Bee Catalogue, 2008) = pseudo (Latin prefix pseudo- "False".) + jatiformis (see jatiformis)

pygmaea, *Trigona* Friese 1933b:[Type locality: Indonesia "Beringin, in der Wäldern oberhalb Mandau (Sumatra), v. Bengkalis. Im Juli 1933"]# (Rasmussen, 2008) = From Ancient Greek *Πυγμαῖος* (Pugmaios, "a member of a race of dwarves")

Mandau River is a river in northern Sumatra, Indonesia.

rhodesica, *Trigona (Meliplebeia) beccarii* subsp. Cockerell and [*Trigona (Meliponula) africana rhodesica* Cockerell 1934a: also *Meliplebeia tanganyikae rhodesica* (Cockerell): Moure 1961]^ (Eardley, 2004) Pertains to Rhodesia + -ica (see suffix – ica)

Rhodesia was an unrecognised state located in southern Africa during the Cold War. From 1965 to 1979, it comprised the region now known as Zimbabwe.

rustica, *Partamona* Pedro & Camargo, 2003 [*Etym.:* from the Latin, *rusticus,* = rustic, referring to the dry fields where the species occurs.]* (Moure's Bee Catalogue, 2008)

saiqui, *Trigona* Friese, H. (1900). [*Etym.:* from the folk name "saiqui"]* (Moure's Bee Catalogue, 2008)

Of or pertaining to Saiqui in Estado do Parana, located in Brazil - 1024 km South of Brasília

salvatoris, *Trigona (Oxytrigona)* Cockerell, T. D. A (1917). [*Typ. Loc.:* San **Salvador**, El **Salvador**.]* (Moure's Bee Catalogue, 2008) = From Latin **salvātor** m (*genitive* **salvātōris**); *third declension* - saviour.

San Salvador (English: *Holy Savior*) is the capital city of the Republic of El Salvador and the capital of the San Salvador department. It is the country's most populated municipality and its most important political, cultural, educational and financial centre.

sarawakensis, *Tetragonula* (Schwarz, 1937) [Typ. loc.: Malaysia "Sarawak: Mt. Dulit,][#] (Rasmussen, 2008) = Refers to Sarawak + **ēnsis** (see suffix **–ensis**).

The State of Sarawak is in eastern Malaysia

savannensis, *Trigona (Tetragona)* Roubik, (1980) [*Typ. Loc.:* Kourou, 30 km NW, French Guiana, Roubik leg.]* (Moure's Bee Catalogue, 2008) = Concerning a **savanna**, + **ēnsis** (see suffix **–ensis**).

Savannah, a grassland ecosystem characterized by the trees being sufficiently widely spaced so that the canopy does not close. The Kourou River empties into the Atlantic Ocean. At the mouth of this river sits the town of Kourou. Long white sand beaches and some rocky outcrops line the town's ocean coast, the riverbank and all points inland, consisting mostly of mangrove and dense tropical rainforest.

seridoensis, *Partamona* Pedro & Camargo, 2003 [*Etym.:* reference to the region, "seridó", where the species occurs.]* (Moure's Bee Catalogue, 2008) = Refers to **Seridó** + **ēnsis** (see suffix **–ensis**);

Serido, a municipality in Paraíba state of northeastern Brazil, is best known for its Sephardic heritage.

setentrionalis, *Trigona (Trigona)* Almeida, M. C. (1995) [*Etym.:* referent to the geographic range of the species.]* (Moure's Bee Catalogue, 2008) = From Venetian **setentrionale** m, f (*m and f plural* **setentrionali**) - northern, north + alis (see suffix –alis)

sooretamae, *Partamona* Pedro & Camargo, 2003 [*Typ. Loc.:* Itarana, Espírito Santo, Brazil. *Etym.:* from the Tupi, "sooretama" = Atlantic forest.]* (Moure's Bee Catalogue, 2008)

Sooretama, a municipality located in the Brazilian state of Espírito Santo.;

silvestris, *Apis* Barrère, P. (1741) [*Nomen nudum* ?]* (Moure's Bee Catalogue, 2008) = From Latin **silvestrī** - dative masculine singular of ***silvestris*** - **silvestris** m, f (*neuter* **silvestre**); *third declension* - Of or pertaining to a forest or wood, rural, wild, living in forests.

silvestriana, *Melipona (Trigona)* Vachal, (1908). [*Typ. Loc.:* Rio Urubu, Goyaz, Brazil; Goyaz, Brazil; Mapiri, Bolivia; Cachabé, Equateur (= Cachabí, Esmeraldas, Ecuador); "Honduras Britannique" (= Belize)]* (Moure's Bee Catalogue, 2008) = Pertains to silvestri (see also **silvestris**) + ana (see suffix **–ana**)

subterranea, *Geotrigona* (Friese, 1901) [*Typ. Loc.:* São Paulo, Brazil, v. Ihering leg.; Pará, Brasil, 25.VIII.1899, 13-17.IX.1899, Ducke leg. **nesting site**: subterranean, folk name: "iruçú-mineiro" Ducke, A. (1916).]* (Moure's Bee Catalogue, 2008) = From Latin **subterrāneus** m (*feminine* **subterrānea**, *neuter* **subterrāneum**); *first/second declension* - underground, subterranean.

Tupi Language URUCU - safe pot or in the ground.

sulina, *Lestrimelitta* Marchi & Melo, 2006 *Typ. Loc.:* Jardim das Américas, Curitiba, Paraná, Brazil, 13.III.2002, G. Melo & P. Marchi. *Etym.:* the name refers to the geographic range of the species]* (Moure's Bee Catalogue, 2008) = sul (from Latin sol – sun) + ina (see suffix **–ina**);

Sulina, Paraná, a town and municipality in the state of Paraná in Southern Brazil.

tanganyikae, *Meliplebeia* var. *rhodesica* (Cockerell): and [Trigona (Meliplebeia) *tanganyikae medionigra* Cockerell: Pooley & Michener 1969]^ (Eardley, 2004) Pertains to Tanganyika with the Latin **ae** terminate.

From 1962 to 1964, the **Republic of Tanganyika** was a sovereign state in East Africa. It was situated between the Indian Ocean and the African Great Lakes of Lake Victoria, Lake Nyasa and Lake Tanganyika.

tarataensis, *Trigona kohli* Schwarz, H. F. (1948). [Nomen nudum]* (Moure's Bee Catalogue, 2008) = Refers to **Tarata** + **ēnsis** (see suffix –**ensis**);

Refers to Tarata District, a district in the Tarata Province, Peru

Sp. 28 Oxytrigona tataira Photo: via Maria Rosa Bernasconi

tataira, *Oxytrigona* (Smith, 1863) [Typ. loc.: Brazil]* (Moure's Bee Catalogue, 2008) = Pertaining to the region named **Tataira,** Paraiba in Brazil. Sp. 28

tellurica, *Geotrigona* Camargo & Moure, 1996 Typ. loc.: Ecuador, Morona-Santiago, Macas, V.1985, G. Onore leg.]* (Moure's Bee Catalogue, 2008) = From English **telluric** - (pertaining to the Earth): earthly, terrestrial, tellurian, terrene

terricola, *Geotrigona* Camargo & Moure, 1996 and also [*Geotrigona lutzi terricola* Camargo & Moure (1996).]* (Moure's Bee Catalogue, 2008) = From Latin *terra* ("earth") + *colō* ("to inhabit") – Latin **terricolus** m (*feminine* **terricola**, *neuter* **terricolum**); *first/second declension* - earth-dwelling; terrestrial

tescorum, *Axestotrigona* (Cockerell) and *Trigona tescorum* Cockerell, 1910 **Syn. n.** (Eardley & Urban, 2010) = **From Latin tescōrum** - genitive plural of *tescum* - **tescum** n (*genitive* **tescī**); *second declension* - (chiefly in the plural) wilderness, wasteland, desert.

tica, *Plebeia* (Wille, 1969) **[***Typ. Loc.:* Playas del Coco, Guanacaste Province, Costa Rica, A. Wille leg.. *Etym.:* from "ticus", a nickname for a Costa Rican.]* (Moure's Bee Catalogue, 2008) = From Portuguese tico m (*feminine* tica *plural* ticos *feminine plural* ticas*; comparable)* - Costa Rican

tobagoensis, *Plebeia* Melo, 2003 [*Typ. Loc.:* Scarborough, **Tobago**, West Indies, 5.IX.2001, *Etym.:* referring to the country of origin, Tobago.]* (Moure's Bee Catalogue, 2008) = Refers Tobago, + ensis (see suffix –ensis)

Tobago is the smaller of the two main islands that make up the Republic of Trinidad and Tobago.

togoensis, *Melipona* Stadelmann, 1895: also [*Trigona erythra var.* ***togoensis*** (Stadelmann): and *Meliponula (Axestotrigona)* ***togoensis*** (Stadelmann)]^ (Eardley, 2004) = Refers to**go** + ensis (see suffix –ensis).

Officially the **Togolese Republic** is a country in West Africa

trinidadensis, *Melipona* Provancher, L. (1888) [Typ. Loc.: Trinidad]* (Moure's Bee Catalogue, 2008) = Refers to Trinidad + ensis (see suffix **–ensis**).

In 1889, the two islands became a single crown colony. **Trinidad** and Tobago obtained self-governance in 1958 and independence from the British Empire in 1962. See also *trinitatis*

trinitatis, *Melipona fasciata* var. Cockerell, T. D. A. (1919).[*Typ. Loc.:* Port of Spain, Trinidad]* (Moure's Bee Catalogue, 2008) = From Latin **trīnitās** f (*genitive* **trīnitātis**); *third declension* - (Ecclesiastical) the Trinity.

Spanish *trinidad* ("trinity"), named after the Holy Trinity by Columbus to fulfill a religious vow.

tropica, *Lestrimelitta* Marchi & Melo, 2006 [Typ. loc.: Farm iguape, Silva Jardim, Rio de Janeiro, Brazil, 8.III.2003, SL Pinto leg. Etym: from the Greek, *tropica,* = living in the tropics]* (Moure's Bee Catalogue, 2008)**.**

Of or pertaining to the tropics, the equatorial region between 23 degrees north and 23 degrees south.

tubiba, *Scaptotrigona* (Smith, 1863) [Typ. loc.: Brazil]* (Moure's Bee Catalogue, 2008) = From Latin **tubus** m (*genitive* **tubī**); *second declension* - tube, pipe; A folk name of the species Scaptotrigona tubiba (Smith, 1863). Sp. 29

Sp. 29 *Scaptotrigona tubiba* Photo: Luiz Carlos Da cunha

From Old Tupi *tubi*, 'bee girl'.Relates to Fazenda Tubiba (Fazenda Tubiba) is a populated place in Estado do Ceara (Ceará), Brazil (South America)

tujumirim, *Scaptotrigona tujumirim* Kerr, et al, (2005). (Nomen nudum) (Moure's Bee Catalogue, 2008) = derived from the folk name 'tujumirim' of the species. From Old Tupi Tuyuba, Tuyuva in São Paulo - SP, Tiúba or Theuba in Bahia,Tuhuba - Tuhuva – Tuyú - yellow bee + mirim, (see **mirim**, "small, tiny").

tumupasae, *Melipona marginata* Schwarz, H. F. (1932). [*Typ. Loc.:* Tumupasa, Bolívia]* (Moure's Bee Catalogue, 2008).

Pertaining to Tumupasa, a region in Bolivia, northeast of Lake Titicaca.;

tungurahuana, *Parapartamona* (Schwarz, 1948) [*Typ. Loc.:* Baños, **Tungurahua**, Ecuador]* (Moure's Bee Catalogue, 2008) = Pertaining **Tungurahua** + ana (see suffix **–ana**);

(/tʊŋgʊˈrɑːwə/); from Quichua *tunguri* (throat), *rahua* (fire): "Throat of Fire" or from Panzaleo) is an active stratovolcano located in the Cordillera of Ecuador. The volcano gives its name to the province of Tungurahua.

tunneli, *Melipona* Pugh 1947: 62: (Nomen nudum)# (Rasmussen, 2008) = From Middle French *tonnelle*. **tunnel** (plural **tunnels**) - A hole in the ground made by an animal, a burrow.

[The statement that it "is different from the other two in appearance as it possesses somewhat yellow spots in the abdomen", however, could make it a *nomen dubium*, although most likely than a junior synonym of *arcifera*]#

urucu, *Trigona* Smith, F. (1868) (Nomen nudum)* (Moure's Bee Catalogue, 2008) = Folk name of this species from Tupi language **urucu** meaning safe pot or in the ground (implying a subterranean nesting species)

uwa, *Paratrigona* Gonzalez & Vélez, 2007 [*Etym.:* name dedicated to **U'wa** people.]* (Moure's Bee Catalogue, 2008).

The **U'wa people** are indigenous people living in the cloud forests of northeastern Colombia. Historically, the U'wa scattered over a homeland that extended across the Venezuela-Colombia border. The U'wa are known to neighbouring indigenous peoples as "the thinking people" or "the people who speak well".

venezuelana, *Trigona* Schwarz, 1948 [*Typ. Loc.:* Lagunita de Aroa, Venezuela, 2000 feet, XII.1910]* (Moure's Bee Catalogue, 2008) = Pertaining to Venezuela + -ana (see suffix **–ana**).;

(officially called the **Bolivarian Republic of Venezuela,** is a country on the northern coast of South America.)

yucatanica, *Melipona (Melipona)* Camargo, Moure & Roubik, 1988 [Typ. loc.: Mérida, **Yucatan**, México, 16.XII.1980, J. C. Acereto]* (Moure's Bee Catalogue, 2008) = Pertaining to the **Yucatan** Peninsula, Mexico + ica (see suffix **–ica**)

sp 10 *Melipona yucatanica* Photo: R. Ayala

yungarum, *Partamona* Pedro & Camargo, 2003 *Etym.:* reference to the **Yungas** mountains, Bolívia, where the species occurs.]* (Moure's Bee Catalogue, 2008) = Pertaining to The **Yungas** + rum (From Latin **rum** meaning "**of**");

A stretch of forest along the eastern slope of the Andes Mountains from Peru, Bolivia and northern Argentina.

Chapter 6

Floral & Faunal Interconnections

The symbiotic relationship of pollinators with plants have been an intriguing phenomenon since ancient history. Flowers and bees are integral components of nature. The stingless bee named *Trigona* ***angophorae*** Cockerell 1912 and *Melipona* **triplaridis** *fulvipes* Cockerell (1925) are clear examples of the pollinator after the flower it predominates. *Partamona* ***epiphytophila*** Pedro & Camargo, 2003 nesting habits shows an affinity to epiphyte roots, so much so its name reflects the nesting substrate of this species.

One can find bees in resin fossil some 80 mya, already showing the existence of this age-old interconnection. This is reflected in Taxon relating to resinous(resin exuding) plants as in *Melipona seminigra* **picea** subsp. Camargo & Roubik (1991 in the Americas Picea forests occur in subtropical mountain systems and humid subtropical forest and *another example is Austroplebeia* **cassiae** (Cockerell, 1910) in Australia.

The predatory and prey syndrome in our ecological life cycle also predates humankind. Insectivorous birds lend their name to this nomenclature, as evident in these examples. *Trigonisca* **ceophloei** (Schwarz, 1938) found in British Guiana and *Trigona* **sarothrum** *(Oxytrigona)* Roubik et al.(1977) endemic to Central American cloud forest.

This life-cycle food chain brings **mutualism** and **commensal** occurrences as relatively eye-opening concepts. Inevitably the naming of species of this abundance of species reflects these concepts, e.g. *Schwarzula* ***coccidophila,*** Camargo & Pedro (2002). indicates the peculiar lifestyle of the species in association with coccids.

Sharing of nesting sites and **Interspecies communication** is communication between different species of animals, plants, fungi or bacteria. Cooperative interspecies communication implies the sharing and understanding of information from two or more species that work towards both species' benefit (mutualism). Most research has found cooperative communication in prey animals whose alarm calls alert conspecifics and heterospecifics. *Partamona* ***gregaria*** Pedro & Camargo, 2003, this name implies some form of interspecies communication as the name's etymology refers to the nests, constructed in aggregations of several nests in the same termitarium.

Meliponines even go to the extent of nesting in their enemies forts, e.g. *Paratrigona* ***myrmecophila*** Moure, 1989, **dare to infiltrate and occupy** active arboreal ant nest: *Camponotus senex* (Formicidae) in Brazil. Another example in South East Asia is *Sundatrigona moorei* (Schwarz, 1937). Although not reflected in its name, the aggressiveness shows when they nest in aerial *Crematogaster* (Formicidae: Hymenoptera) nests.

Sp. 30 *Melipona quadrifasciata anthidioides* Photo: via Mari Rosa Bernasconi

Meliponine Taxa with Flora & Fauna Relationship

Taxa and Epithets with Flora & Fauna connections

angophorae, *Trigona* **Cockerell** 1912. [Typ. Loc.[: Australia "Sydney, New South Wales, Dec. 1, 1910" Floral record: *Angophora*][#] (Rasmussen, 2008)

Relates to ***Angophora***, a genus of flowering plants in the myrtle family. Angophora is endemic to Australia (Wikipedia, Angophora, 2013)

anthidioides, *Melipona* Lepeletier, 1836 syn. to *Melipona (Melipona) quadrifasciata anthidioides* Lepeletier, 1836 [*Typ. Loc. :* Minas Gerais, Brasil]* also refers to [*Melipona* ***anthidioides*** *integrior* Cockerell, 1919 *Typ. Loc.:* Sapucay, Paraguay]* (Moure's Bee Catalogue, 2008) = anthi (see **anthi** "small bird") + di (see **di** "diety") + oides (see suffix **–oides** "resembling") – resembling a small bird diety. **(anthi - From Latin anthus** m (*genitive* **anthī**); *second declension* - a small bird, possibly the yellow wagtail.). Sp. 30

Correndera Pipit [Anthus correndera] The Correndera Pipit (Anthus correndera) is a species of bird in the Motacillidae(Yellow wagtail) family. It is found in Argentina, Bolivia, **southern Brazil**, Chile, the Falkland Islands, **Paraguay**, Peru, and Uruguay. Its natural habitats are temperate grassland, subtropic or tropical high-altitude grassland, and pastureland. (Wikipedia, Correndera Pipit, 2013).

cassiae, *Austroplebeia* (Cockerell, 1910) [Floral record: *Cassia*][#] (Rasmussen, 2008) = From Latin **cassia** f (*genitive* **cassiae**); *first declension* - cinnamon; Cassia is also the English common name of some species in the genus *Cinnamomum*;

Relating to ***Cassia***, a genus of flowering plants in the legume family

ceophloei, *Trigonisca* (Schwarz, 1938) [*Typ. Loc.* : Kartabo, British Guiana, 28.VII.1920, from the stomach of a woodpecker, **Ceophloeus** lineatus (Aves, Picidae).]* (Moure's Bee Catalogue, 2008)

Figure 4 *Dryocopus lineatus* (Wikipedia, Lineated Woodpecker 2013)

***Ceophloeus** [Dryocopus] lineatus lineatus* (Linnaeus, 1766) – From Costa, Colombia and east to Trinidad, the **Guianas** and Brazil (Maranhão, Bahia) (São Paulo)., Peru, northern Paraguay. Bill dark. Body mass is 186–228 g. Total length is 34 cm, Tail 145 mm. Wing from 195mm to 200mm, and peak at the height of 36 mm. (Wikipedia, Lineated Woodpecker, 2013)

citriperda, *Melipona* Illiger, K. (1806) (Nomen nudum) (Moure's Bee Catalogue, 2008) = citri Latin – English inflected **citri**, "lemon" + perda From Old Portuguese *perda*, from Latin *perdita*, feminine of *perditus* ("lost"), the perfect passive participle of *perdō* ("I lose").;

Relates to a species of fruit flies *Ceratitis* ***citriperda*** Macleay, 1829 syn. ***Ceratitis capitata***, the **Mediterranean fruit fly**, or **medfly** for short, is a species of fruit fly capable of causing extensive damage to a wide range of fruit crops. It is native to the Mediterranean area but has spread invasively to many parts of the world, including Australasia and North and South America. (Wikipedia, Ceratitis citriperda, 2014)

Figure 5 Coccidae (Wikipedia, Scale Insects 2013)

coccidophila, *Schwarzula* Camargo & Pedro (2002). [*Etym.*: reference to the peculiar lifestyle of the species (association with coccids)]* (Moure's Bee Catalogue, 2008) = cocci(see **cocci**) + do (see **-do**) + phila (see suffix **phila**);

These bees are the only known species that have, within their own nest, a permanent source of carbohydrates ('honeydew' from the Coccids(scale insects)) (Wikipedia, Scale Insects, 2013), and only pollen is collected at flowers (Camargo and Pedro 2002).

cuculina, *Melipona* Müller, (1875). (Nomen nudum)* (Moure's Bee Catalogue, 2008) =' cuculi (see **cuculi**) + -ina (see suffix **–ina**); Fritz Muller made a description, "a bee format different from the others who lived among the colonies of some species as a parasite, calling it cuculina Melipona" (Ribeiro et al. 2006) ** (Krug, 2010) **cuculi** - From Latin **cuculus** m (*genitive* **cucūlī**); *second declension* - cuckoo (bird).

The **cuckoos** are a family of **Cuculidae** - Many species are brood parasites, laying their eggs in the nests of other species.

epiphytophila, *Partamona* Pedro & Camargo, 2003 [*Etym.:* from the Greek, *epi* + *phytón* + *philos*, indicating the nesting substrate: among roots of epiphytes.]* (Moure's Bee Catalogue, 2008) = from *ἐπί* (epi, "on top of") + *φυτόν* (phyton, "plant"), via -phyte. - **epiphyte** + phila ("a liking or tendency");

A plant that grows on another using it as physical support but neither obtaining nutrients from it nor causing it any damage if also offering no benefit; an air plant.

gregaria, *Partamona* Pedro & Camargo, 2003 *Etym.:* from the Latin, *gregarius*, referring to the nests, which are constructed in aggregations of several nests in the same termitarium.]* (Moure's Bee Catalogue, 2008)

isopterophila, *Aparatrigona* Schwarz, 1934* (Moure's Bee Catalogue, 2008) = From Latin - iso(see prefix **iso-**) + ptera (see **ptera**) **isoptero** - A taxonomic order within the subclass Pterygota - termites). + phila (see suffix **-phila**).

[describing a species that have a liking for termites] recently, were classified at the taxonomic rank of order **Isoptera**, but are now accepted as the infraorder **Isoptera**, they are for practical purposes unrelated to the ants.

limao, *Lestrimelitta* (Smith, 1863)* (Moure's Bee Catalogue, 2008) = **From Portuguese limão** m (*plural* **limões**) - lemon (citrus fruit) [cleptobiosis, behaviour, robbing, scent glands, lemon-like odor,]^` (Moure et al., 1958).

A feature, well known in the case of *L. limao,* is the rapid domination of victims by the invaders. Apparently, they accomplish this by overwhelming the colony odour of the robbed hive with their own limelike mandibular gland secretion^^ (Bohart, 1970)

mimetica, *Melipona (Michmelia)* Cockerell, 1914 also [Melipona **mimetica** Cockerell, (1914). *Etym.:* the specific name refers to the perfect superficial resemblance to *Centris tarsata* Smith.]* (Moure's Bee Catalogue, 2008) = From Ancient Greek *μιμητικός* (mīmētikos, "imitative"). **Mimetic** - Exhibiting mimesis - (biology) mimicry.

mosquito, *Plebeia* Smith, 1863- also referring to [*Melipona* Ducke, 1916, *Trigona* Rau, 1933 and *Trigona* ***mosquito*** *domiciliorum* Schwarz, 1934 (junior subjective synonym) .]* (Moure's Bee Catalogue, 2008) - Of or pertaining to mosquito(**mosquito** (*plural* **mosquitos** *or* **mosquitoes**).

A small flying insect of the family *Culicidae*, known for biting and sucking blood, leaving an itching bump on the skin.

mulata, *Partamona* Moure, in Camargo, 1980 (Moure's Bee Catalogue, 2008) = Spanish **mulata** f (*plural* **mulatas**, *masculine* **mulato**) - From Spanish or Portuguese *mulato* ("of mixed breed, young mule"), from *mulo* ("mule"), from Latin *mulus* ("mule").

Used as an allusion to the hybrid origin of mules. - **Mulatto** (*plural* **mulattoes** *or* **mulattos**) - (often derogatory) A person of mixed African and Caucasian descent. (Wikipedia, Mulatto, 2014)

musarum, *Trigona* Cockerell, 1917 also [*Partamona* ***musarum*** (Cockerell, 1917) *Type locality:* Philadelfia **Banana** R., Costa Rica, F. Knab. *Etym.:* from the botanical name *Musa*, for **banana**.]* (Moure's Bee Catalogue, 2008) = Latin **mūsārum** - genitive plural of *mūsa* - **Musa** – plant genus of bananas + rum(see **rum**)

muscaria, *Trigona* Spinola, (1840). (Nomen nudum)* (Moure's Bee Catalogue, 2008) = From Latin **muscārius** m (*feminine* **muscāria**, *neuter* **muscārium**); *first/second declension* - Of or pertaining to flies, fly

myrmecophila, *Paratrigona* Moure, 1989 [Etym.: nesting site: active arboreal ant nest: *Camponotus senex* (Formicidae)]* (Moure's Bee Catalogue, 2008) = **From Latin myrmecophile** (*plural* **myrmecophiles**).

An organism, especially an insect, that lives in close association with or shares a nest with a species of ant. See also suffix **-phila.**

oyani, *Trigona* Darchen, 1971 Syn. n. *Meliponula (Axestotrigona) oyani* (Darchen)^ (Eardley, 2004) = Oyani is an animal of the genus Poiana in the family Viverridae. They are like a lot and linisang'i of Asia (*Prionodon*)

Oyani is small (30 cm body and tail of over 30 cm) but thin with short legs. Their colour is yellow with black spots (spotted, striped) and their tail have black ringlets. (Wikipedia, Oyani, 2013) Figure 6

Figure 6 Oyani (Wikipedia, Oyani 2013)

phenax, *Melipona (Melipona)* Cockerell, 1919 (Moure's Bee Catalogue, 2008) = refers to plant genus Phenax (family of Asian Ghostweed)

phrynostoma, *Plebeia* Moure, 2004 - [*Etym.:* from de Greek, *phrynós* - toad and *stoma* - mouth, referring to the shape of the nest entrance.]* (Moure's Bee Catalogue, 2008) = From Ancient Greek *Φρύνη* (Phrunē), from *φρύνη* (phrunē, "toad") - Translingual ***Phryne*** - (obsolete) *Bufo(Latin word for toad.),* certain frogs + stoma(see **stoma**)

picea, *Melipona seminigra* subsp. Camargo & Roubik (1991). (Nomen nudum; *Melipona seminigra picea* is an undescribed subspecies of *Melipona seminigra* Friese, 1903)* (Moure's Bee Catalogue, 2008) = Latin **picea** f (*genitive* piceae); *first declension* - spruce, pitch pine

poecilochroa, *Plebeia* Moure & Camargo, 1993 (Moure's Bee Catalogue, 2008) - English Prefix **'pœcilo-'** - Alternative spelling of poikilo- From Ancient Greek *ποικίλος* (poikílos, "spotted, embroidered"). Prefix **poikilo-** (sciences) variegated or variable + chroa(Greek χρωμη **chroa** – chrome, colour);

Related to Poecilochroa, a spider genus of the family araneomorfas Gnaphosidae. It is located in Eurasia, Africa and South America. (Wikipedia, Gnaphosidae, 2013)

prosopiformis, *Paratrigona* Gribodo, 1893 – also referring to [*Melipona* Gribodo, 1893 by original designation]* (Moure's Bee Catalogue, 2008) pro(see prefix **pro-**) + sopi(see **sopi**) + formis(see **formis**);

Prosopis is a genus of flowering plants in the pea family, Fabaceae. It contains around 45 species of spiny trees and shrubs found in subtropical and tropical regions of the Americas, Africa, Western, and South Asia. They often thrive in arid soil and are resistant to drought, on occasion developing extremely deep root systems. Their wood is usually hard, dense and durable. Their fruits are pods and may contain large amounts of sugar. The generic name means "burdock" in late Latin and originated in Greek (Wikipedia, Prosopis, 2014)

pseudomusarum, *Partamona (Partamona)* Camargo, (1980)* **(Moure's Bee Catalogue, 2008)** - pseudo(see prefix **pseudo-)** + musarum(see **musarum**)

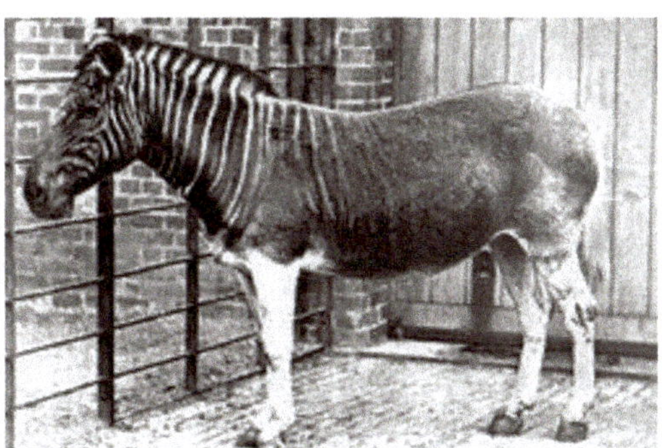
Figure 7 Quagga (Wikipedia, Quagga 2014)

pulla, *Melipona* **Illiger** 1806: 158: [Nomen nudum. The name is unavailable for taxonomic purposes, as no description or definition of the taxon was provided (ICZN, art. 12.1); Indonesia, Sumatra; *Melipona pulla* **Dalla**

Torre 1896][#] (Rasmussen, 2008) = Latin - ablative feminine singular of pullus - **pullus** (*genitive* pullī); m, second declension - young (of animals), chick, chicken

quagga, *Trigona* Strand, 1911^ (Eardley, 2004) - Of or pertaining to the **quagga** (/ˈkwəˈxə/) (*Equus quagga quagga*) is an extinct subspecies of the plains zebra that lived in South Africa.

It was long thought to be a distinct species, but recent genetic studies have shown it to be the plains zebra's southernmost subspecies. It is considered particularly close to Burchell's zebra. Its name is derived from the plains zebra's call, which sounds like "kwa-ha-ha" (Wikipedia, Quagga, 2014)

rhodopta, *Trigona (Trigonisca)* Wille & Michener (1973) (Nomen nudum) under *Trigonisca discolor* (Wille, 1965)*[&] (Jarau & Barth, 2008) - occurs in the Golfo Dulce region, Costa Rica; possibly relates to **rhodophyta** "Red Algae"**.**

The red coralline algae occur as red tides in the Golfo Dulce fiord. (Wikipedia, Red Algae, 2014)

sarothrum, *Trigona (Oxytrigona)* Roubik et al.(1977)* (Moure's Bee Catalogue, 2008) - Pertaining to Sarothrura(flufftails) + -rum(see **rum**).

Possibly relates to **Sarothrura**, the taxonomic genus of the **flufftails**, placed in the family ***Sarothruridae***, small birds related to the rails. (Wikipedia, Sarothrura, 2013)\

solani, *Melipona (Michmelia)* Cockerell, 1912 **[**Cockerell, T. D. A. (1912) Descriptions and records of Bees - XLVI. *Ann. Mag. Nat. Hist.* (as *Melipona solani*, **flower record**: *Solanum*)]* (Moure's Bee Catalogue, 2008).

sp 11 *Melipona solani* Photo: John S. Ascher

Solanum is a large and diverse genus of flowering plants, including two food crops of the highest economic importance, the potato and the tomato. (Wikipedia, Solanum, 2014)

topiorum, *Trigona* Cockerell, 1910. [Also refers to *Meliplebeia topiorum* (Cockerell)]* (Moure's Bee Catalogue, 2008) - From Latin **topio** – landscape + rum (From Latin **rum** meaning "**of**");

Pertaining to **Topi,** a highly social and fast antelope (Wikipedia, Topi, 2014)

triplaridis, *Melipona fulvipes* Cockerell, T. D. A. (1925). [*Typ. Loc.:* Balboa, Canal Zone, Panamá, 26-27.March, W. M. Wheeler leg., at flowers of Triplaris cumingiana.]* (Moure's Bee Catalogue, 2008).

Triplaris is a genus of plants in the family Polygonaceae. The species are variously distributed in the Americas. Some species are used for lumber. (Wikipedia, Triplaris, 2013)

vulpina, *Melipona* Cockerell, 1919* (Moure's Bee Catalogue, 2008) = From earlier Latin *volpēs*, from Proto-Italic **wolpi*, Proto-Indo-European (red) fox**;** Refers to the red fox colour; From Latin **vulpīnus** m (*feminine* **vulpīna**, *neuter* **vulpīnum**); *first/second declension* - vulpine- Pertaining to a fox. The **red fox** (*Vulpes vulpes*) is the largest of the true foxes and the most geographically spread member of the Carnivora (Wikipedia, Red Fox, 2014)

Figure 8 Red Fox (Wikipedia, Red Fox 2014)

zebra, *Trigona* Strand, 1911a: [The specific epithet zebra was attributed to Friese by Strand (1911a), but Friese's description appeared only in 1912. Meade-Waldo (1913) synonymised *Melipona denoiti* and *Trigona clypeata,* whereas Cockerell (1920) considered zebra to be a synonym of *M. denoiti.*]^ (Eardley, 2004) = From Portuguese *zebra*, *zebro* ("zebra"), from Old Portuguese *enzebro*, *ezebra*, *azebra* ("wild ass"), from earlier *cebrario* (882), *ezebrario* (897), from Vulgar Latin **eciferus*, from Latin *equiferus* ("wild horse") (Pliny).

English **zebra** (*plural* **zebras**) - Any of three species of genus *Equus*: *E. grevyi*, *E. quagga* or *E. zebra*, all with black and white stripes and native to Africa (Wikipedia, Zebra, 2014)

zexminiae, *Cephalotrigona* (Cockerell, 1912**)** and [*Trigona zexmeniae* Cockerell, 1912: flower records: *Zexmenia virgulta*]* (Moure's Bee Catalogue, 2008).

Relating to ***Zexmenia***, a genus of flowering plants in the daisy family, Asteraceae (Wikipedia, Zexmenia, 2013).

sp 12 *Cephalotrigona zexmenia* Photo: R. Ayala

Floral Rec. 9 *Solanum avuculare* (Wikimedia, Poroporo 2014)

Floral Rec. 9 *Angophora* (Wikimedia, *Angophora* 2014)

Floral Rec. 9 *Cassia* (Wikimedia, *Cassia brewsteri* 2009)

Floral Rec. 9 *Citrus limon* (Wikimedia 2014)

Floral Rec. 9 *Musa sp.* (Wikimedia, Fleurs_de_bananier 2014)

Floral Rec. 9 *Picea pungens* (Wikimedia, *Picea pungens* 2014)

Floral Rec. 9 *Prosopis pallida* (Wikimedia, *Prosopis pallida* 2014)

Floral Rec. 9 *Triplaris* (Wikimedia, *Triplaris americana* 2014)

Floral Rec. 9 *Zexmenia virgulata* (Meuseum n.d.)

Legend (Flora & Faunal Section):

** A comunidade de abelhas (Hymenoptera – Apoidea) de duas areas de interesse biológico e histórico em Santa Catarina Cristiane Krug 2010

^` Moure, J. S., P. Nogueira-Neto & W. E. Kerr (1958). Evolutionary problems among Meliponinae (Hymenoptera, Apoidea).

^^ Bohart, George E., "The Evolution of Parasitism Among Bees" (1970).

*+ Smith, G.F.; Walker, Colin and Figueiredo, E. (2011). What's in a name: an update on epithets used in Aloe L. (Asphodelaceae). Bradleya, 29 pp. 179–181.

*& Stefan Jarau & Friedrich G. Barth (2008). Stingless bees of the Golfo Dulce region, Costa Rica (Hymenoptera, Apidae, Apinae, Meliponini)

Web Reference:

Christopher Taylor - Nature and Wildlife Photography - Correndera Pipit [Anthus correndera] http://www.kiwifoto.com/galleries/birds

Lineated Woodpecker

http://en.wikipedia.org/wiki/Dryocopus_lineatus

Chapter 7

Anatomy attributes and Color Descriptive Epithets

This section deals with epithets that pertains to the Meliponine anatomical breakdown. The anatomical parts' size and colour description can be used to indicate prominent features of the bee.

Latin terms are used to name the anatomy/

Pes = feet

Crus = legs

Ceps = head

Cornis = antennae

Pennis = wings

Ventris or gastra = belly

How to use this section: e.g. Synonymy

1. Trigona spinipes (Fabricius, 1793) Sp. 31

Apis spinipes Fabricius, J. C. (1793).

SUBSEQUENT DESIGNATION: Moure, J. S. (1960). Notes on the types of the Neotropical bees described by Fabricius (Hymenoptera: Apoidea). *Stud. Entomol. (Rio de Janeiro)* **3 (1-4):** 97-160 [155] .

Apis ruficrus Latreille, P. A. (1804). (junior subjective synonym)

Melipona citriperda Illiger, K. (1806). (*Nomen nudum*, see Latreille, 1811: 293) .

Trigona muscaria Spinola, M. (1840). (*Nomen nudum*)

Trigona luteipes Smith, F. (1854). (junior subjective synonym).

Trigona ruficrus concolor Gribodo, J. (1894). (junior subjective synonym) .

Trigona ruficrus rufoides Strand, E. (1910). (junior subjective synonym).

Spinipes (spiny feet), ruficrus (reddish brown legs), citriperda & muscaria (fly-like), luteipes (saffron yellow feet)

Sp. 31 Trigona spinipes Image source: http://en.wikipedia.org/wiki/Trigona_spinipes

2. *Trigona hyalinata* (Lepeletier, 1836) Sp. 32

Melipona atratula Illiger, K. (1806). (*Nomen nudum*) *Melipona argentata* Lepeletier, A. L. M. (1836). (junior subjective synonym, according to personal notes of Pe. Moure on the type of *Melipona argentata* Lepeletier)

Melipona hyalinata Lepeletier, A. L. M. (1836). By inference, the designation of the lectotype of *Trigona hyalinata* can be attributed to Schwarz (1948: 236-237). Moure, in 1958, in litt.

Trigona hyalina Spinola, M. (1840). (*Nomen nudum*)

Hyalina[ta] (glassy), ruficrus (reddish brown legs), argentata

Sp. 32 *Trigona hyalinata* Image Source: http://www.ame-rio.org/2010/04/abelha-caipira.html

Trigona flavipennis Friese, H. (1900). (junior subjective synonym)

Trigona friesei Ihering, H. (1903). (junior subjective synonym).

Melipona ruficrus flavidipennis Ducke, A. (1916). (*unnecessary nom. nov.* for *Trigona flavipennis* Friese, 1900).

Anatomy and colour attribute epithets

aethiops, *Trigona* Müller, F. (1921) (Nomen nudum)* (Moure's Bee Catalogue, 2008) = From Ancient Greek *Aἰθίοψ* (Aithiops), from *αἴθω* (aithō, "burn") + *ὤψ* (ōps, "face"). Compare *Aethiops* and *Ethiopia*. **aethiops** - A specific name of several black animals

affabra, *Scaptotrigona* (Moure, 1989) and [*Sakagamilla affabra* Moure, (1989)]* (Moure's Bee Catalogue, 2008) = From Latin **affabrē** (*not comparable*) - ingeniously, skilfully

albipennis, *Trigona albipennis* Almeida, 1995* (Moure's Bee Catalogue, 2008) = Latin **albipennis** m, f (*neuter* albipenne); *third declension* - white-winged, white-feathered

albofasciata, *Trigona beccarii* var. Friese, 1916; Alternative combination - *Trigona (Meliponula) albofasciata* (Friese): Cockerell 1934a ^ (Eardley, 2004) = albo (From Latin *albus* ("white")) + fasciata (see **fasciata**); white banded

alisfumatis, *Trigona Mexicana* subsp. Sapper, K. T. (1935)* (Moure's Bee Catalogue, 2008) = alis (Latin "wings") + **fūmātis** From Latin *fūmus*. - *supine* fūmātum - smoke, steam, fume.; having smoky wings.

ambusta, *Trigona* Cockerell 1918 [Typ. loc.: Malaysia/Singapore "Sandakan and Singapore"][#] (Rasmussen, 2008) = Latin - scorched, burnt; syn. for *Geniotrigona thoracica* (Smith, 1857) Sp. 33*Error! Reference source not found.*

Sp. 33 *Geniotrigona thoracica* Photo: Ahmad Zaini

angulata, *Melipona* Spinola, M. (1840)* (Moure's Bee Catalogue, 2008) = From Latin **angulātus** m (*feminine* angulāta, *neuter* angulātum); *first/second declension* - cornered, angular

angustula, *Tetragonisca* (Latreille, 1811) and syn. [*Trigona* ***angustula*** Latreille, P. A. (1811) also *Melipona* ***angustula*** Illiger, K. (1806). (Nomen nudum)]* (Moure's Bee Catalogue, 2008) = From Latin *angustus*. **-Angust** - Narrow; strait; contracted+ ula(see suffix **–ula**); describes the narrow contracted abdomen of this species.

***apicalis**, Tetrigona* (Smith, 1857)[#] (Rasmussen, 2008) = Latin m *and* f, **apicale** n; *third declension* - apical(of or connected with the apex) + alis (Latin "wings"); having apical wings. Sp. 34

***apiformis**, Melipona (Michmelia)* (Du Buysson & Marshall, 1892)* (Moure's Bee Catalogue, 2008) = New Latin; from *apis* ("bee") + *forma* ("form, shape"); having the form of a bee

Sp. 34 *Tetrigona apicalis* Photo Hisham Ismail

***argentata**, Melipona* Lepeletier, A. L. M. (1836)* (Moure's Bee Catalogue, 2008) = From Latin **argentātus** m (*feminine* argentāta, *neuter* argentātum); *first/second declension* - silvered, silvery

***argyrea**, Scaura* Cockerell, 1912* (Moure's Bee Catalogue, 2008) = From the Ancient Greek *ἄργῠρος* (arguros, "silver", "money"). English Prefix '**argyr-**' -silver + **ea** (Latin "that way or that side")

***atratula**, Melipona* Illiger, K. (1806)* (Moure's Bee Catalogue, 2008) = From Latin atratus - clothed in black (for mourning), atrata ("darkened, blackened") + ula (see suffix **–ula)**

***atricornis**, Platytrigona* (Smith, 1865)[#] (Rasmussen, 2008) = From Latin *ater* ("black") + cornis (antenna); having blackish antenna.

***atripes**, Tetragonilla* (Smith, 1857)[#] (Rasmussen, 2008) = From Latin *ater* ("black") + *pēs* ("foot"). **atripēs** m, f, n, (*genitive* **atripedis**); *third declension* - black-footed. Sp. 35

Sp. 35 *Tertagonilla atripes* Photo: Ahmad Zaini

***atroalba**, Trigona* Ducke, A. (1908)* (Moure's Bee Catalogue, 2008) = Latin atro ("dark") + alba(Latin **alba -** a precious white stone, the pearl).; a darker hue of pearly white

***atrolutea**, Partamona* Pedro & Camargo (2003)* (Moure's Bee Catalogue, 2008). = Latin atro ("dark") + lutea (Latin "yellow"); dark yellow

auripennis, *Partamona* Pedro & Camargo, 2003 [Etym.: from the Latin, *aureus* = golden, + *penna* = wing.]* (Moure's Bee Catalogue, 2008) having gold coloured wings

basalis, *Trigona* Smith, F. (1854)[#] (Rasmussen, 2008) = Latin -m *and* f, **basale** n; *third declension* - basal, having a pedestal or foot

basimaculata, *Melipona* Bingham 1903* (Moure's Bee Catalogue, 2008) = **basi** - Latin- **basis** - A pedestal, foot, base; + maculate (Latin –"spotted or stained"); having stained feet

bicolour, *Melipona* Lepeletier, (1836) and syn. of subsp. [*Melipona (Eomelipona) bicolor bicolor* Lepeletier, 1836]* (Moure's Bee Catalogue, 2008) = Latin **bicolor** - Of two colours or having two colours. Sp. 36

Sp. 36 *Melipona bicolor* Photo: Ivan Romero

bidentata, *Trigonisca* Albuquerque & Camargo, 2007 [Etym.: from the Latin, *bi* + *dentatus*, referring to the bidentate mandible.]* (Moure's Bee Catalogue, 2008) = From **(*bi-*** (Latin "two") + ***dentatus*** (Latin -Having tooth-like projections that are themselves dentate.))

bilineata, *Partamona* Say, 1837* (Moure's Bee Catalogue, 2008) = From *bi* + *lineatus Latin* **bilīneātus** m (*feminine* bilīneāta, *neuter* bilīneātum); *first/second declension* - doubly lined

bipartita, *Melipona* Lepeletier (1836).* (Moure's Bee Catalogue, 2008) = From Latin **bipartītus** m (*feminine* bipartīta, *neuter* bipartītum); *first/second declension* - bipartite (divided in two parts)

Sp. 37 *Scaptotrigona bipunctata* Photo: Luis Emilio Velaquez

bipunctata, *Scaptotrigona* (Lepeletier, 1836) and subsp. [*Scaptotrigona bipunctata polysticta* Moure, (1950).]* (Moure's Bee Catalogue, 2008) = From Latin **bipunctātus** - doubly punctuated or pointed; bipunctate Sp. 37

bismarkiana, *Trigona bismarkiana* Tadauchi *et al.* 1998 [Nomen nudum] # (Rasmussen, 2008) = bismark (pigment *Bismark brown* when used as a biological stain) + iana (see suffix **–iana**)

bitika, *Liotrigona* Brooks & Michener, 1988 = From Malagasy(Madagascan) – **bitika** means **diminutive**; Smallest bee in Madagascar. ^ (Eardley, 2004) *ITIS Bees: World Bee Checklist*. Ruggiero M. (project leader), Ascher J. et al., 2009-09-28 (ITIS, 2009) http://www.catalogueoflife.org/annual-checklist/2011/details/species/id/6971938

bivea, *Meliwillea* Roubik, et al. (1997). [*Etym.:* from the Latin, *bivius*, meaning to have two paths (by a *lapsus*, the name was incorrectly spelt as *bivea* by the authors)]* (Moure's Bee Catalogue, 2008)

bomboides, *Melipona* Dominique, 1898 * (Moure's Bee Catalogue, 2008) = Latin *bombo* **bombo** m (*plural* bombos) - (colloquial) the bump (stomach of a pregnant woman) + oides(Latin **-oides** "resembling")

brachychaeta, *Melipona (Michmelia)* Moure, 1950* (Moure's Bee Catalogue, 2008) = **Brachy-** Prefix indicating short + chaeta (resembling chitin.).

brevis, *Trigona* Friese 1925 [Nomen nudum][#] (Rasmussen, 2008), also Hym. *Trigona* ***brevis*** Friese, H. (1925) [Nomen nudum]* (Moure's Bee Catalogue, 2008) = Latin **brevis** - short, low, shallow; According to John Ascher and Claus Rasmussen (*in litt.*) the type of *Trigona brevis*, deposited in the AMNH, with label of Brasil/Jundiahy, is indeed a *Tetragonula* of the *fuscobalteata* (Cameron, 1908) group from Indo-Malayan region.).

breviceps, *Trigona* Cockerell 1919b syn. of *Heterotrigona itama* (Cockerell, 1918) [#] (Rasmussen, 2008) = Latin **brevis** - short, low, shallow + ceps (Latin - "head"); having a shallow head

brevipilosa, *Parapartamona* Schwarz, 1948* (Moure's Bee Catalogue, 2008) = brevi (Latin **brevis** - short, low, shallow) + pilosa(Latin **pilosa** - hairy, shaggy.); having short hairs

caerulea, *Mourella* (Friese, 1900)* (Moure's Bee Catalogue, 2008) = **From Latin caeruleus** m (*feminine* caerulea, *neuter* caeruleum); *first/second declension* - dark coloured, dark blue or dark green, cerulean, azure

canifrons, *Lophotrigona* (Smith, 1857)# (Rasmussen, 2008) = cani (Latin **cani** - (of hair) grey) + frons (Latin **frons** - forehead.); having a grey-haired frons (the upper anterior portion of the head of an insect, above or behind the clypeus)

capitata, *Cephalotrigona* Smith, 1854* (Moure's Bee Catalogue, 2008) = Latin – capitāta nominative feminine singular of ***capitātus***- **capitātus** m (*feminine* capitāta, *neuter* capitātum); *first/second declension* - having, or forming a head

captiosa, *Melipona (Michmelia)* Moure, 1962* (Moure's Bee Catalogue, 2008) = From Latin **captiōsus** m (*feminine* captiōsa, *neuter* captiōsum); *first/second declension* - captious, harmful

carbonaria, *Tetragonula* (Smith, 1854)# (Rasmussen, 2008) = Latin (*genitive* carbōnāriae); f, *first declension* - furnace for making charcoal. Sp. 38

castanea, *Melipona* Bingham 1903# (Rasmussen, 2008) = Relating to the chestnut tree or the **colour** of chestnut. The name *Castanea* is probably derived from the old name for the sweet chestnut in Ancient Greek

Sp. 38 *Tetragonula carbonaria* Photo: Robert Lutrell

catabolonota, *Paratrigona* Camargo & Moure, 1994 [Etym.: from the Greek: *katabollo* = truncated, plus *noto* = back, referring to the anterior margin of the mesoscutum]* (Moure's Bee Catalogue, 2008)

catira, *Lestrimelitta* Gonzalez & Griswold, 2012 [Etym.: the specific epithet is a word used by Venezuelan people to describe a blonde or red-headed person; this word is used in reference to the reddish colouration of this species.]* (Moure's Bee Catalogue, 2008)

chrysura, *Melipona* Cockerell, 1918^ (Eardley, 2004) = from greek χρυσ *chrys*(gold) + ουρα *oura*(tail) – golden tail or gold-coloured rear body part.

ciliata, *Lestrimelitta* Marchi & Melo, 2006 [Etym.: from the Latin, *ciliatus*]* (Moure's Bee Catalogue, 2008) = From Latin **ciliātus** m (*feminine* ciliāta, *neuter* ciliātum); *first/second declension* - Having cilia (eyelid); Usage notes: Used almost exclusively as a taxonomic epithet.

cilipes, *Trigona* Fabricius, 1804* (Moure's Bee Catalogue, 2008) = From Latin cili - chisel + pes (Latin "foot"); having chiselled feet

cincta, *Austroplebeia* (Mocsáry in Friese, 1898) also *Trigona* ***cincta*** subsp. *percinta* nov. (Rasmussen, 2008) = Latin - **cinctus** m (*feminine* cincta, *neuter* cinctum); *first/second declension* - wreathed, crowned, girded, bordered,

clavicornis, *Dolichotrigona* Camargo & Pedro, 2005 [Etym: from the Latin, *clava* + *cornu* = horn, antenna, referring to the clavate form of the antennal scapus.]* (Moure's Bee Catalogue, 2008)

clavipes, *Tetragona* Fabricius, 1804* (Moure's Bee Catalogue, 2008) = clavi(see **clavi**) + pes(see **pes**); From *clava* ("club, cudgel") + *pēs* ("foot"). - **clavipēs** m, f, n (*genitive* clavipedis); *third declension* - club-footed. Sp. 39

Sp. 39 *Tetragona clavipes* Photo: via Maria Rosa Bernasconi

clypearis, *Trigona* Dominique, 1898* (Moure's Bee Catalogue, 2008) and [*Tetragonula* ***clypearis*** (Friese, 1909) also syn. subsp. *Trigona laeviceps* variety ***clypearis*** Friese 1909("1908")][#] (Rasmussen, 2008) = Latin - **clipeus** (*genitive* clipeī); m, *second declension* - A small round shield + -aris (Latin -indicating a relationship or pertaining to) - pertaining to the clypeus.

(notes to *Trigona clypearis* - The synonymy proposed here was based on photos of the specimens deposited in NHNN, sent by Gilles Mahé and Alain Pauly to Claus Rasmussen.)* (Moure's Bee Catalogue, 2008)

clypeata, Trigona Friese, 1909 and [*Plebeina* ***clypeata*** (Friese) also *Trigona* ***clypeata*** var. *zebra* Strand]^ (Eardley, 2004) = From Latin **clipeātus** m (*feminine* clipeāta, *neuter* clipeātum); *first/second declension* - shielded (armed with a shield)

combinata, *Partamona* Pedro & Camargo, 2003 [*Etym.:* from the Latin, *combinatus*, referring to the labrum colour, which can be black or yellow.]* (Moure's Bee Catalogue, 2008)

compressa, *Trigona* Latreille, (1811) * (Moure's Bee Catalogue, 2008) = Latin compressa - constricted/narrow/pressed together

compressipes, *Melipona (Melikerria)* (Fabricius, 1804) and also [*Melipona* ***compressipes*** *afinis* Posey, D. A. (1983).]* (Moure's Bee Catalogue, 2008) = English **compress** - (transitive) To make smaller; to press or squeeze together, or to make something occupy a smaller space or volume. + pes (Latin foot). Sp. 40

Sp. 40 *Melipona compressipes* Photo: via Mari Rosa Bernasconi

compsa, *Paratrigona* Camargo & Moure [Etym.: from the Greek, *kompsa* = charming, elegant.]* (Moure's Bee Catalogue, 2008) = *From Latin* **compsi -** (in general) adorn, deck, dress, ornament.

concolor, *Trigona* Gribodo, 1895* (Moure's Bee Catalogue, 2008) = Latin **concolor** m, f, n (*genitive* concolōris); *third declension* - of the same colour or uniform colour throughout

corvina, *Trigona* Cockerell, 1913* (Moure's Bee Catalogue, 2008) = From Latin *corvus* ("crow") + *-īnus*. Latin **corvīnus** m (*feminine* corvīna, *neuter* corvīnum); *first/second declension* - corvine; of or pertaining to crows or ravens (dark colours).

crassicornis, *Paratrigona* Camargo & Moure, 1994 [Etym.: from the Latin, *crassus* = thick, plus *cornus* = antenna.]* (Moure's Bee Catalogue, 2008) = This species has thick solid horns/antennae

crassipes, *Trigona* (Fabricius, 1793) and syn. *Apis crassipes* Fabricius, J. C. (1793)* (Moure's Bee Catalogue, 2008) = crassi(Latin "thick, solid") + pes (Latin "foot"); having thick solid feet.

crispula, *Leurotrigona* Pedro & Camargo, 2009 [Etym.: From the Latin, crispus, relating to the sinuous, curly hairs of the posterior parapenicillum.]* (Moure's Bee Catalogue,

2008) = From Latin **crispus** - curly; crimped (of hair), tremulous + ula(Latin suffix – **ula** indicating a diminutive form)

cubiceps, *Cleptotrigona* (Friese), and syn. *Lestrimelitta (Trigona) cubiceps* Friese, 1912^ (Eardley, 2004) = From Latin **cubus** m (*genitive* cubī); *second declension* - A mass, quantity + ceps (Latin - "head"); having a mass of head

curtula, *Melipona* Illiger, K. (1806) (Nomen nudum)* (Moure's Bee Catalogue, 2008) = Latin **curtus** *m* (*feminine* curta, *neuter* curtum); *first/second declension* - shortened, mutilated, incomplete + -ula (see suffix –ula)

cyanescens, *Trigona* Friese, H. (1911)* (Moure's Bee Catalogue, 2008) = From Latin **cȳaneus** m (*feminine* cȳanea, *neuter* cȳaneum); *first/second declension* - deep or dark blue (colour) + scens (Latin **scens** – **"**scene, appearance")

delimbata, *Trigona nebulata* var. Strand, 1914^ (Eardley, 2004) = delimb - English To cut the limbs from. + -ata (see suffix –ata)

depilis, *Scaptotrigona* Moure, 1942* (Moure's Bee Catalogue, 2008) = Latin dē-pīlis - pilus, - without hair From Latin **depilare** *present active* **dēpilō**, *present infinitive* dēpilāre, *perfect active* dēpilāvī, *supine* dēpilātum - pull out the hair; depilate

discolour, *Trigonisca* Wille, 1965* (Moure's Bee Catalogue, 2008) = Latin **discolor** m, f, n (*genitive* discoloris); *third declension* - having a different colour or variegated

dissecta, *Tetragona* Moure, 2000* (Moure's Bee Catalogue, 2008) = From Latin *dissecare* ("to cut asunder, cut up"), from *dis-* ("asunder") + *secare* ("to cut"); **dissect** - (transitive, anatomy, surgery) To separate muscles, organs, and so on without cutting into them or disrupting their architecture.

dolichogaster, *Melipona (Tetragona)* (Kohl): Vachal 1903a and *Trigona* ***dolichogaster*** Kohl, 1894 (Eardley & Urban, 2010) = From Ancient Greek *δολιχός*. English Prefix **dolicho-** = long + **gaster** f (*genitive* gasteris); *third declension* - belly: having a long belly

dorsalis*, Tetragona* (Smith, 1854)* (Moure's Bee Catalogue, 2008) = **Latin dorsālis (**From *dorsum* ("the back")) m, f (*neuter* dorsāle); *third declension* - dorsal; of or pertaining to the back

Sp. 41 Melipona eburnea Photo: Claus Rasmussen

eburnea*, Melipona (Michmelia)* Friese, 1900* (Moure's Bee Catalogue, 2008) = **From Latin eburneus** m (*feminine* eburnea, *neuter* eburneum); *first/second declension* - of ivory, white as ivory Sp. 41 Melipona eburnea Photo: Claus Rasmussen

eburneiventer*, Cephalotrigona* (Schwarz, 1948)* (Moure's Bee Catalogue, 2008) = eburnei (Latin **eburnei – "ivory")** + venter (Latin **venter** - belly, abdomen)

elegantula*, Trigona* Müller, F. (1921) *Incertae sedis** (Moure's Bee Catalogue, 2008) = From a diminutive of *elegans* Latin **elegantulus** m *first/second declension* - small and elegant + ula(Latin suffix **–ula** indicating a diminutive form)

elongata*, Trigona* Lepeletier & Serville, (1828)* (Moure's Bee Catalogue, 2008) = From Latin **ēlongātus** m (*feminine* ēlongāta, *neuter* ēlongātum); *first/second declension* - elongated, having been stretched out.

emerina*, Plebeia* Friese, 1908* (Moure's Bee Catalogue, 2008) = English emery (raw corundum used for sanding) + ina (From Latin -inus -Of or pertaining to)

emerinoides*, Trigona* Silvestri, F. (1902)* (Moure's Bee Catalogue, 2008) = emerin (see **emerina)** + oides(see suffix **–oides**),; resembling emerina. see also ***emerina***

Sp. 42 *Heterotrigoan erythrogastra* Photo by Author

erythra*, Axestotrigona* (Schletterer) and syn. [*Melipona (Trigona)* ***erythra*** (Schletterer, 1891]^ (Eardley, 2004) = From Greek: Ερυθρά *Erythra* (red);

The **Erythraean Sea** (Greek: Ερυθρά Θάλασσα, *Erythra Thalassa*; "Red Sea") is the name in ancient cartography for a body of water located between the Horn and the Arabian peninsula.

erythrogastra, *Heterotrigona* Cameron, 1902# (Rasmussen, 2008) = From Latin **erythrogaster** m (*feminine* **erythrogastra**, *neuter* **erythrogastrum**); *first/second declension* -having a red belly see also *erythra.* Sp. 42

erythrostoma, *Trigona* **Cameron** 1908 syn. of *Tetragonula fuscobalteata* (Cameron, 1908)# (Rasmussen, 2008) = (erythro - Prefix: used to form scientific terms meaning. red. or showing a relationship to red) + (Stoma - From Ancient Greek *στόμα* (stoma, "mouth"). (zoology) A mouthlike opening

Sp. 43 *Melipona fasciculata* Photo Giorgio Venturieri

eutaeniata, *Paratrigona* Camargo & Moure, 1994 [Etym.: from the Greek, *eu* = very, quite a lot, plus *tainia* = belt, strip.]* (Moure's Bee Catalogue, 2008) = Latin eu- ("well") + taeniata (from Latin **taeniātus** – be-ribboned (to trim with ribbon) + -ata(From Latin Suffix '-**ata'** - Added to nouns to form adjectives Having, -ed.) well ribboned or belted.

euxanthospila, Paratrigona Camargo & Moure, 1994 [Etym.: from the Greek, *eu* + *xanthos* + *spilos* = well marked with yellow.]* (Moure's Bee Catalogue, 2008)

fasciata, *Melipona (Michmelia)* Latreille, 1811* (Moure's Bee Catalogue, 2008) = **Latin-** nominative feminine singular of fasciātus=**fasciātus** m (*feminine* **fasciāta**, *neuter* **fasciātum**); *first/second declension* - bundled, banded

fasciculata, *Melipona (Melikerria)* Smith, 1854* (Moure's Bee Catalogue, 2008) = **From Latin fasciculātus** m (*feminine* **fasciculāta**, *neuter* **fasciculātum**); *first/second declension* - fasciculate, fasciculated - a diminutive of fascis.**fascicle** (plural fascicles) - A bundle or cluster. (anatomy): A bundle of skeletal muscle fibres surrounded by connective tissue. Sp. 43

Sp. 44 *Melipona favosa* Photo: via Mari rosa Bernasconi

favosa, *Melipona (Melipona)* (Fabricius, 1798)* (Moure's Bee Catalogue, 2008) = Latin *favus* honeycomb. English **favose** - (botany) honeycombed. Sp. 44

femorata, *Cephalotrigona* Smith, 1854* (Moure's Bee Catalogue, 2008) = Latin - **femorāta** - having the form of a thigh see also ***femoralis***

femoralis, *Paratrigona* Camargo & Moure, 1994 [Etym.: from the Latin, referring to the special shape of the femur.]* (Moure's Bee Catalogue, 2008) = Latin **femorālis** m, f (*neuter* femorāle); *third declension* - having the form of a thigh

ferrea, *Trigona* Cockerell 1929b# (Rasmussen, 2008) = Latin -ablative feminine singular of *ferreus* - **ferreus** m (*feminine* ferrea, *neuter* ferreum); *first/second declension* - like or pertaining to iron or colour of rust

ferricauda, *Trigona* Cockerell, 1917* (Moure's Bee Catalogue, 2008) = Latin **ferri** - from ferrum n (*genitive* **ferrī**); ***second declension*** - **iron** + cauda Latin **cauda** f (*genitive* caudae); *first declension* - A tail (*of an animal*).; having an iron or iron-rust coloured tail.

ferruginea, *Meliponula (Axestotrigona)* (Lepeletier) comb.n., *Melipona (Trigona) ferruginea* Lepeletier, 1841 ^ (Eardley, 2004) = From Latin ferrum - ferru m (plural ferri) - iron + ginea – Latin- responsible; **ferru ginea**= Latin phrase meaning Iron image

fimbriata, *Homotrigona* (Smith, 1857)# (Rasmussen, 2008) = Latin **fimbriātus** m (*feminine* **fimbriāta**, *neuter* **fimbriātum**); *first/second declension* - fringed, fibrous. Sp. 45

Sp. 45 *Homotrigona fimbriata* Photo: Ahmad Zaini

flaveola, *Oxytrigona* (Friese, 1900) and *Melipona (Trigona) flaveola* Spinola, M. (1840). (Nomen nudum)* (Moure's Bee Catalogue, 2008) = From Latin **flaveolus** m (*feminine* flaveola, *neuter* flaveolum); *first/second declension* - yellowish

flavicans, *Trigonisca* (Moure, 1950) and syn. *Hypotrigona (Trigonisca) graeffei* ***flavicans*** Moure, J. S. (1950)* (Moure's Bee Catalogue, 2008) = From *flāveō - Latin* **flavicans** - yellowing (becoming yellow)

flavicornis, *Frieseomelitta* (Fabricius, 1798) and syn. *Apis flavicornis* Fabricius, J. C. (1798)* (Moure's Bee Catalogue, 2008) = Latin flavi (yellow) + cornis (antenna); **flavicornis** m, f (*neuter* fulvicorne); *third declension* - having yellowish antennas

flavidipennis, *Melipona ruficrus* subsp. Ducke, A. (1916). (*unnecessary nom. nov.* for *Trigona flavipennis* Friese, 1900)* (Moure's Bee Catalogue, 2008) = Latin **flāvidī** - nominative masculine plural of *flāvidus* -**flāvidus** m (*feminine* flāvida, *neuter* flāvidum); *first/second declension* - yellowish + pennis ("wings"); having yellowish wings.

flavipennis, *Trigona* Friese, H. (1900). and *Melipona* ***flavipennis*** Smith, F. (1854)* (Moure's Bee Catalogue, 2008) = Latin flavi(yellow) + pennis(wing); **flavipennis** m, f (*neuter* flavipenne); *third declension* - having yellow wings

flavistigma, *Trigona* Cameron 1902 syn. to *Homotrigona fimbriata* (Smith, 1857)[#] (Rasmussen, 2008) = Latin flavi (yellow) + Latin **stigma** *n* (*genitive* **stigmatis**); *third declension* - brand (burned mark, scar); having a yellow burned mark

flavisetis, *Nannotrigona (Scaptotrigona) postica* Absy, et al. (1984)* (Moure's Bee Catalogue, 2008) = Latin flavi (yellow)+ Latin setis (bristles or hair); having yellow bristles

flaviventris, *Platytrigona* (Friese, 1909)[#] (Rasmussen, 2008) = Latin flavi (yellow) + *venter* ("belly"). *third declension* - having a yellow belly

flavocincta, *Plebeia* (Cockerell, 1912)* (Moure's Bee Catalogue, 2008) = flavo (Latin "yellow or blond") + cincta (Latin "girded or bordered"); yellow girded or crowned.

flavofasciata, *Melipona scutellaris* subsp. Ducke, A. (1916)* (Moure's Bee Catalogue, 2008) = flavo (Latin "yellow or blond") + fasciata (Latin "banded"); having yellow/golden bands.

flavolineata, *Melipona (Michmelia)* Friese, 1900* (Moure's Bee Catalogue, 2008) = **From Latin flavolineātus** m (*feminine* flavolineāta, *neuter* flavolineātum); *first/second declension* - marked with yellow lines. Sp. 46

Sp. 46 *Melipona flovolineata* Photo: Giorgio venturieri

flavornata, *Schwarziana* Cruz-Landim, & Mota, (1990)* (Moure's Bee Catalogue, 2008) = flavo (Latin "yellow or blond") + ornata (Latin – ornate decoration) ; Having yellowish decorations.

flavoscutellata, *Plebeia* Camargo, (1988)* (Moure's Bee Catalogue, 2008) = flavo(see **flavo**) + scutella(Latin "shield") + ata(Latin – Of or pertaining to); having a yellow scutellum

foetida, *Trigona trombeta* subsp. Müller, F. (1921)* (Moure's Bee Catalogue, 2008) = From Latin foetidus m (*feminine* **foetida**, *neuter* **foetidum**); *first/second declension* - bad-smelling, foul, stinking

frontalis, *Trigona* Friese, H. (1911)* (Moure's Bee Catalogue, 2008) = Latin frontalis m, f (*neuter* **frontale**); ***third declension*** - **frontal** - (anatomy) Of or relating to the frontal plane.

fuliginosa, *Melipona (Michmelia)* Lepeletier, 1836* (Moure's Bee Catalogue, 2008) = **From Latin fūlīginōsus** m (*feminine* fūlīginōsa, *neuter* fūlīginōsum); *first/second declension* - sooty

fulvatra, *Geotrigona* Camargo & Moure, 1996* (Moure's Bee Catalogue, 2008) = New Latin; from *fulvus* ("reddish yellow") + atra(Latin atro "dark"); dark reddish yellow (amber colored)

fulvicutis, *Scaptotrigona* Moure, 1964) and syn.[*Nannotrigona (Scaptotrigona) fulvicutis* Moure, (1964)]* (Moure's Bee Catalogue, 2008) = New Latin; from *fulvus* ("tawny") + cutis ("living skin"); having a reddish-yellow cuticle

fulvipes, *Melipona* Guérin-Méneville, F. E. (1844).and subsp. [*Melipona fulvipes triplaridis* Cockerell, (1925)]* (Moure's Bee Catalogue, 2008) = fulvi (Latin "amber coloured") + pes (Latin "foot"); having amber-coloured feet.

Sp. 47 *Trigona fulviventris Photo:* via Mari Rosa Bernasconi

fulviventris, *Trigona* Guérin, 1844* (Moure's Bee Catalogue, 2008) = fulvi(see **fulvi**) + ventris(see **ventris**); **fulviventris** (New Latin; from *fulvus* ("tawny") + *venter* ("belly"). *third declension* - having a tawny belly. Sp. 47

fulvohirta, *Geotrigona* (Friese, 1900)* (Moure's Bee Catalogue, 2008) = New Latin; from *fulvus* ("tawny") + hirta(Latin hirtus "hairy, shaggy"); having tawny or amber coloured hairs

fulvomarginata, *Trigona* Cockerell 1919c syn. to ***Lepidotrigona terminata*** **(Smith, 1878)#** (Rasmussen, 2008) = New Latin; from *fulvus* ("tawny") + Latin - **marginātus** (border or margin) - having a tawny border or margin

fulvopilosa, *Plebeia (Plebeia)* Ayala, (1999) [Etym.: colour of the pilosity]* (Moure's Bee Catalogue, 2008) = From Latin fulvus "**reddish-yellow"** + pilosa(Latin "hairs"); having amber coloured shaggy hairs

fulvopilosella, *Trigona* Cameron 1908 syn to *Lepidotrigona nitidiventris* (Smith, 1857)# (Rasmussen, 2008) = From Latin fulvus "**reddish-yellow"** + pilosa(Latin "hairs") - having amber coloured shaggy hairs + -ella (From Latin suffix **–ellus** —indicates a diminutive form**)**

fumata, *Parapartamona* Moure, 1995* (Moure's Bee Catalogue, 2008) = **From Latin fūmātus** m (*feminine* fūmāta, *neuter* fūmātum); *first/second declension* - (New Latin) smoky (coloured)

fumipennis, *Trigona fumipennis* Lutz, F. E. (1924). (Nomen nudum)* (Moure's Bee Catalogue, 2008) = Latin **fumipennis** m, f (*neuter* fumipenne); *third declension* - having smoky (coloured) feathers/wings

furva, *Lisotrigona* Engel, 2000[#] (Rasmussen, 2008) = Latin- **furvus** m (*feminine* **furva**, *neuter* **furvum**); *first/second declension* - dark, dusky, gloomy, swarthy, black

fuscata, *Melipona (Melipona)* Lepeletier, 1836* (Moure's Bee Catalogue, 2008) = From Latin **fuscātus** m (*feminine* fuscāta, *neuter* fuscātum); *first/second declension* - darkened

fuscibasis, *Tetragonilla* (Cockerell, 1920)[#] (Rasmussen, 2008) = fusci (see **fusci**) + **basis** - Latin - An underlying condition or circumstance. Sp. 48

Sp. 48 *Tetragonilla fuscibasis* Photo: Ahmad Zaini

fuscipes, *Melipona* Friese, H. (1900)* (Moure's Bee Catalogue, 2008) = fusci (see **fusci**) + pes("foot"); having dark feet

fuscobalteata, *Tetragonula* (Cameron, 1908)* (Moure's Bee Catalogue, 2008) = fusco (Latin – darken) + balteata (Latin **balteatus** - furnished with a girdle or belt) Sp. 49

Sp. 49 *Tetragonula fuscobalteata* Photo: Ahmad Zaini

fuscopilosa, *Melipona (Michmelia)* Moure & Kerr, 1950 and syn. [*Melipona eburnea* ***fuscopilosa*** Moure & Kerr (1950)]* (Moure's Bee Catalogue, 2008) = fusco(Latin – "darken") + pilosa (Latin – "hairs"); having dark dusky hairs.

fuscipennis, *Trigona* Friese, 1900* (Moure's Bee Catalogue, 2008) = fusci(From Latin fusco – "darken") + pennis (Latin pennis – "wings"); having dark dusky wings

genalis, *Papuatrigona* (Friese, 1909)[#] (Rasmussen, 2008) = Latin *gena* ("cheek") + Latin lis - (anatomy) Of or relating to the cheeks

geniculata, *Melipona* Illiger, (1806)* (Moure's Bee Catalogue, 2008) = From Latin **geniculātus** m (*feminine* geniculāta, *neuter* geniculātum); *first/second declension* - knotty (full of knots)

glabella, *Paratrigona* Camargo & Moure, 1994 [Etym.: from the Latin, *glabella* = without hairs.]* (Moure's Bee Catalogue, 2008)

glaberrima, *Lestrimelitta* Oliveira & Marchi, 2005 also *Trigona amalthea* ***glaberrima*** Friese, (1913). [Etym.: Latin, *glabro* (glabrous) + suffix, *-érrimo*.]* (Moure's Bee Catalogue, 2008) = Latin **glaberrima** - very hairless or smooth

glabrata, *Lestrimelitta* Moure & Camargo, 1989 and [*Paratrigona glabrata* Moure, (1990)]* (Moure's Bee Catalogue, 2008) = **Latin glabrātus** m (*feminine* glabrāta, *neuter* glabrātum); *first/second declension* - somewhat smooth; glabrate

gracilis, *Leurotrigona* Pedro & Camargo, 2009 [Etym.: From the Latin, gracilis, slender, delicate, slight.]* (Moure's Bee Catalogue, 2008)

grandis, *Melipona (Melikerria)* Guérin, 1844* (Moure's Bee Catalogue, 2008) = **Latin grandis** m, f (*neuter* grande); *third declension* - large, great, grand, tall, lofty ***grandipennis,*** *Partamona* (Schwarz, 1951)* (Moure's Bee Catalogue, 2008) = grand(see also ***grandis***) + pennis(Latin – "wings"); having grandeur wings. Sp. 50

Sp. 50 Melipona grandis Photo: Claus Rasmussen

grandipennis, *Partamona* (Schwarz, 1951)* = grand(see also ***grandis***) + pennis(Latin – "wings"); having grandeur wings.

haematoptera, *Odontotrigona* (Cockerell, 1919)[#] (Rasmussen, 2008) = haemato – (English – of blood or blood colour) + ptera – from Ancient Greek.- **pterus** - winged Sp. 51

hemileuca, *Trigona* Cockerell 1929 syn to *Tetrigona apicalis* (Smith, 1857) [#] (Rasmussen, 2008) = hemi- From Ancient Greek prefix *ἡμι-* (hēmi-, "half"), + leuca – white; referring to the half white wings of this species.

Sp. 51 *Odontotrigona haematoptera* Photo: Claus Rasmussen

hirsuticornis, *Celetrigona* Camargo & Pedro, 2009 [Etym.: From the Latin, *hirsutus*, shaggy, bristly, plus *cornis*, antenna, referring to the long, erect bristles on the flagellum.]* (Moure's Bee Catalogue, 2008)

hirtella, *Melipona* Spinola, (1840)* (Moure's Bee Catalogue, 2008) = hirta (Latin – "hairy, shaggy") + -ella (From Latin suffix **–ellus** —indicates a diminutive form)

hirticornis, *Trigonisca* Albuquerque & Camargo, 2007 [*Etym.:* from the Latin, *hirtus* (= hairy) + *cornus* (= antenna).]* (Moure's Bee Catalogue, 2008) having a hairy antenna

hispida, *Trigona* Marianno, (1911)* (Moure's Bee Catalogue, 2008) = From Latin **hispidus** m (*feminine* hispida, *neuter* hispidum); *first/second declension* - rough, fuzzy, hairy, bristly, prickly

hyalina, *Trigona* Spinola, M. (1840) (Nomen nudum)* (Moure's Bee Catalogue, 2008) = From Latin **hyalinus** m (*feminine* hyalina, *neuter* hyalinum); *first/second declension* - glassy, hyaline, glassy-green

hyalinata, *Melipona* Lepeletier, (1836)* (Moure's Bee Catalogue, 2008) = Latin *hyalinus*, from late Ancient Greek *ὑάλινος*, from *ὕαλος* ("glass"). - Glassy, transparent, amorphous. + -ata(Added to nouns to form adjectives Having, -ed.) see also ***hyaline.*** Sp. 32

ignis, *Oxytrigona* Camargo, 1984* (Moure's Bee Catalogue, 2008) = Latin **ignis** m (*genitive* **ignis**); *third declension* – Pertaining to ignite or fire

Sp. 52 *Melipona illota* Photo: Claus Rasmussen

illota, *Melipona (Michmelia) illota* Cockerell, 1919 and[*Melipona scutellaris illota* Cockerell, (1919)]* (Moure's Bee Catalogue, 2008) = Latin **illota** – unwashed, unkempt(dishevelled; untidy; dirty; not kept up) Sp. 52

illustris, *Melipona (Eomelipona)* Schwarz, 1932 and subsp. [*Melipona marginata illustris* Schwarz, H. F. (1932)]* (Moure's Bee Catalogue, 2008) = Latin **illustris** m, f (*neuter* illustre); *third declension* - bright, shining, brilliant, lustrous

imberbis, *Parapartamona* Moure, 1995* (Moure's Bee Catalogue, 2008) = Latin **imberbis** m, f (*neuter* imberbe); *third declension* - beardless

impunctata, *Aparatrigona* (Ducke, 1916)* (Moure's Bee Catalogue, 2008) = im- (From Latin in- negation prefix) + punctata (From Latin **punctātus** - punctuated, pointed) = (*anatomy*) not punctate (that is, **not** marked by spots, dots, points, or punctures).

incīsa, *Geniotrigona* (Sakagami & Inoue, 1989)# (Rasmussen, 2008) = Latin **incīsus** m (*feminine* **incīsa**, *neuter* **incīsum**); *first/second declension* - cut up, dissected, carved, engraved, having been engraved Sp. 53

Sp. 53 *Geniotrigona incisa* Photo: via Paimin Ponijan.

infuscata, *Trigona nebulata* Cockerell, 1934a* (Moure's Bee Catalogue, 2008) = Latin īnfuscātus - darken or dusky in colour or obscured

itama, *Heterotrigona* (Cockerell, 1918)# (Rasmussen, 2008) = From Proto-Malayo-Polynesian *(ma-)qitəm*. **Itam**-black (absorbing all light) Sp. 54

Sp. 54 *Heterotigona itama* Photo Rosafandi Rosli

iridipennis, *Tetragonula* (Smith, 1854)# (Rasmussen, 2008) = iridi (Latin iridare – "To make iridescent") + pennis (Latin –"wings"); having iridescent wings

lacteifasciata, *Geniotrigona* (Cameron, 1902)# (Rasmussen, 2008) = lactei (Latin – "milky") + fasciata ("banded"); having milky bands. Sp. 55

Sp. 55 *Genitrigona lacteifasciata* Photo: Morilee Mahar

lacteipennis, *Trigona* Friese, 1900* (Moure's Bee Catalogue, 2008) = lactei (Latin – "milky") + pennis (Latin – "wings"); having milky wings

***laeviceps**, Tetragonula* (Smith, 1857)[#] (Rasmussen, 2008) = Latin **laevi** - genitive neuter singular of laevus = **laevus** m (*feminine* **laeva**, *neuter* **laevum**); *first/second declension* - left; on the left side., (by extension) awkward, foolish + ceps (Latin - "head"); having an awkward head. Sp. 56

Sp. 56 Tetragonula laeviceps Photo: Acai Ibrahim

***latebalteata**, Lepidotrigona* (Cameron, 1902)[#] (Rasmussen, 2008) = Latin **lātē** (*comparative* **lātius**, *superlative* **lātissimē**) broadly, widely + balteata (Latin **balteatus** - furnished with a girdle or belt); having a broad belt.

***latigenalis**, Trigona (Tetragonula)* Sakagami 1978 syn. to *Tetragonula reepeni* (Friese, 1918) [#] (Rasmussen, 2008) = lati (Latin **lati** - broad) + gena ("cheek"); Having a broad gena

***latipēs**, Lepidotrigona* Friese, 1900)[#] (Rasmussen, 2008) = lati (Latin **lati** - broad) + *pēs* ("foot"). Latin -m, f, n, (*genitive* **latipedis**); *third declension* - wide-footed, wide-legged. Sp. 57

Sp. 57 Lepidotrigona latipes Photo: Azmi Bin Awang Azmi

***latitarsis**, Scaura* (Friese, 1900)* (Moure's Bee Catalogue, 2008) = lati (Latin **lati** - broad) + tarsis (From Latin **tarsus** - flat part of the foot).; having broad tarsus.

***lautipes**, Melipona variegatipes* Cockerell, (1919)* (Moure's Bee Catalogue, 2008) = lauti (From Latin **lautus** - elegant, luxurious) + *pēs* ("foot").; having elegant feet.

***leucogastra**, Geotrigona* (Cockerell, 1914)* (Moure's Bee Catalogue, 2008) = From Latin **leucogaster** m (*feminine* leucogastra, *neuter* leucogastrum); *first/second declension* - having a white belly

***ligata**, Trigona* Say, (1837)* (Moure's Bee Catalogue, 2008) = From Latin **ligātus** m (*feminine* ligāta, *neuter* ligātum); *first/second declension* - bandaged, wrapped, having been bandaged.

lineata, *Paratrigona* var. *glabella* Camargo, & Moure (1994)* (Moure's Bee Catalogue, 2008) = From Latin **līneātus** m (*feminine* līneāta, *neuter* līneātum); *first/second declension* - lined

liturata, *Melipona* Spinola, M. (1840)* (Moure's Bee Catalogue, 2008) = From Latin *lituratus*, past participle of *liturare* to erase, from *litura* a blur. Engliah liturate - (*zoology*) Having indistinct spots, paler at the margins.

lineatifrons, *Paratrigona* (Schwarz, 1938) Etym: the name refers to the yellow markings on the frons.]* (Moure's Bee Catalogue, 2008)

longicollis, *Trigona* Friese, (1903)* (Moure's Bee Catalogue, 2008) = New Latin; from *longus* ("long") + *collum* ("neck, throat"). - **longicollis** *m, f* (*neuter* **longicolle**); *third declension* - having a long neck

longicornis, *Celetrigona* (Friese, 1903)* (Moure's Bee Catalogue, 2008) = From Latin *longus* ("long") + *cornus* ("horn, antenna"). **longicornis** m, f (*neuter* longicorne); *third declension* - long-antenna

longicrus, *Melipona* Dalla Torre (1896)* (Moure's Bee Catalogue, 2008) = Latin - From ***longus*** ("far, long") + Latin **crūs** n (*genitive* crūris); *third declension* - ("a leg") - long-legged

longipes, *Trigona* Smith, (1863)* (Moure's Bee Catalogue, 2008) = Latin - From *longus* ("far, long") + *pēs* ("foot") - **longipēs** m, f, n (*genitive* longipedis); *third declension* - long-footed,

longula, *Scaura* (Lepeletier, 1836)* (Moure's Bee Catalogue, 2008) = **From Latin longulus(**Diminutive of *longus* ("far, long").) m (*feminine* longula, *neuter* longulum); *first/second declension* - Rather long, far or lengthy; large in extent.

longitarsis, *Dolichotrigona* (Ducke, 1916)* (Moure's Bee Catalogue, 2008) = New Latin; from *longus* ("long") + *tarsus* ("foot"). **longitarsis** m, f (*neuter* longitarse); *third declension* - long-tarsus

lunulata, *Melipona (Melipona)* Friese, 1900* (Moure's Bee Catalogue, 2008) = From Latin **lūnulātus** m (botany, zoology) Resembling a small crescent.

lophocoryphe, *Paratrigona* Moure, 1963* (Moure's Bee Catalogue, 2008) = lopho- [from Greek **lophos** "crest"] + Latin coryphé - the chief, A leader of a clan or tribe, A leader of a group

lurida, *Ptilotrigona* (Smith, 1854)* (Moure's Bee Catalogue, 2008) = From Latin **lūridus** m (*feminine* lūrida, *neuter* lūridum); *first/second declension* - pale yellow, wan, sallow, lurid

lutea, *Trigona mocsaryi* subsp. Friese, (1903)* (Moure's Bee Catalogue, 2008) and *Homotrigona lutea* (Bingham, 1897)# (Rasmussen, 2008) = **Latin lūteus** m (*feminine* **lūtea**, *neuter* **lūteum**); *first/second declension* - saffron yellow. Sp. 58

Sp. 58 *Homotrigona lutea* Photo : Azmand Rosland

luteicornis, *Trigona cupira* Pedro & Camargo, (2003)* (Moure's Bee Catalogue, 2008) = Latin lutei (saffron yellow) + cornis ("horn, antennae"); having saffron yellow antennae.

luteipennis, *Scaptotrigona* (Friese, 1902)* (Moure's Bee Catalogue, 2008) = Latin lutei (saffron yellow) + pennis ("wing, feather").; having saffron yellow wings.

luteipes, *Trigona* Smith, (1854)* (Moure's Bee Catalogue, 2008) = Latin lutei (saffron yellow) + pes ("feet").; having saffron-yellow feet

luteiventris, *Trigona* Friese 1909("1908") syn. to *Heterotrigona erythrogastra* (Cameron, 1902)# (Rasmussen, 2008) = Latin lutei (saffron yellow) + venter ("belly, abdomen").; having saffron yellow belly or abdomen.

luteofasciata, *Melipona* Müller, (1921)* (Moure's Bee Catalogue, 2008) = From luteo (saffron yellow) + fasciata ("banded"); saffron yellow-banded

malaris, *Plebeia* Moure, 1962* (Moure's Bee Catalogue, 2008) = From Latin *mala* ("jaw, cheek-bone"). + -aris (Latin suffix **–aris** "Pertaining to") - Pertaining to the jaw (malar).

marginata, *Melipona* Lepeletier, (1836)* (Moure's Bee Catalogue, 2008) = Latin - **marginātus** m (*feminine* margināta, *neuter* marginātum); *first/second declension* - having a border or margin

medionigra, *Trigona (Meliponula) africana* subsp. Cockerell, 1934a^ (Eardley, 2004) = Latin medio ("halve") + nigra ("black"); being half black.

mediorufa, *Oxytrigona* (Cockerell, 1913) and subsp. *Trigona flaveola mediorufa* Cockerell, (1913)* (Moure's Bee Catalogue, 2008) = Latin medio ("halve") + rufa ("red, reddish"); being mediocre red

melanaspis, *Paratrigona* Camargo & Moure, 1994 [Etym.: from the Greek, *melanos* = black, + *aspis* = shield, referring to the clypeus dark, black.]* (Moure's Bee Catalogue, 2008)

melanica, *Plebeia* Ayala, 1999 [*Etym.*: from the Greek, *melanus*.]* (Moure's Bee Catalogue, 2008) = Italian melanica f - **melanic** - Of, relating to, or exhibiting melanism.; Dark pigmentation

melanocera, *Nannotrigona* Schwarz, 1938* (Moure's Bee Catalogue, 2008) = (Latin melano- dark or black) + (cera – beeswax, honeycomb); having a dark honeycomb

Sp. 59 *Tetragonula melanocephala* Photo: Syafrizal Fachmy

melanocephala, *Tetragonula* (Gribodo, 1893)[#] (Rasmussen, 2008) = (Latin melano- dark or black) + **cephala** = New Latin *cephala*, from Ancient Greek κεφαλή (kephalē, "head")head, **cephalate** (zoology) - Having a dark head. Sp. 59

melanoleuca, *Geotrigona* Moure & Camargo (1992) also *Tetrigona melanoleuca* (Cockerell, 1929)* (Moure's Bee Catalogue, 2008) = From melano (dark or black) + leuca(white); referring to the white-tipped black wings. Sp. 71

melanotricha, *Trigona* Cockerell 1918 syn. to *Homotrigona anamitica* (Friese, 1909)[#] (Rasmussen, 2008) = (Latin melano- dark or black) + tricha - **Ancient Greek *θρίξ*** (thríks, "hair"),

melanopleura, *Melipona fasciata* subsp. Cockerell, (1919)* (Moure's Bee Catalogue, 2008) = (Latin melano- dark or black) + pleura – Latin (ribs or sides)

Sp. 60 *Melipona melanoventer* Photo: Giorgio Venturieri

melanoventer, *Melipona (Michmelia)* Schwarz, 1932* (Moure's Bee Catalogue, 2008) = (Latin melano- dark or black) + venter (Latin belly, abdomen) Sp. 60

melina, *Tetragonula* (Gribodo, 1893)# (Rasmussen, 2008) = Latin-From melinus ("colour of honey") Sp. 61

Sp. 61 *Tetragonula melina Photo: Syafrizal Fachmy*

mellaria, *Nannotrigona* (Smith, 1862)* (Moure's Bee Catalogue, 2008) = From Latin **mellārius** - Of or pertaining to honey.

mellea, *Trigona* Smith, (1863)* (Moure's Bee Catalogue, 2008) = From Latin **melleus** - resembling honey, as sweet as honey; honey-sweet,

mellicolor, *Oxytrigona* (Packard, 1869)* (Moure's Bee Catalogue, 2008) = From Latin **mellīs** - honey-water; pertaining to the colour of honey-water

mellipes, *Tetragonula* (Friese, 1898)# (Rasmussen, 2008) = From Latin **mellīs** - honey-water + pes (feet); having honey-water colored feet.

minima, *Plebeia* (Gribodo, 1893)* (Moure's Bee Catalogue, 2008) = From Latin **minima** - nominative feminine singular of *minimus* - **minimus** (*superlative of* parvus) - least, smallest

minor, *Tetragonula* (Sakagami, 1978)# (Rasmussen, 2008) and *Partamona (Nogueirapis) minor* Moure &. Camargo (1982)* (Moure's Bee Catalogue, 2008) = From Latin *minor* ("rather small")

minuta, *Nannotrigona* (Lepeletier, 1836)* (Moure's Bee Catalogue, 2008) = From Latin **minūtus** m (*feminine* minūta, *neuter* minūtum); *first/second declension* - diminished, having been diminished

mirim, *Trigona* Müller, F. (1874) also *Melipona inhati mirim*, Schwarz, 1948* (Moure's Bee Catalogue, 2008) = From Old Tupi.**mirim** (*mf*) - tiny, little, wee; Inhati mirim is the folk name of this species

mixta, *Melipona* Lepeletier, (1836)* (Moure's Bee Catalogue, 2008) = From Latin **mixtus** - **"mixed,** having been mixed"; Having two or more separate aspects.

monodonta, *Lestrimelitta* Moure & Camargo, 1989* (Moure's Bee Catalogue, 2008) = mono (one) + **ódónta-** odonto- From Ancient Greek *ὀδούς* (odous, "tooth"); having one denticle on its mandible.

nana, *Lestrimelitta* Melo, 2003 and *Melipona* ***nana*** Illiger, K. (1806). (Nomen nudum)* (Moure's Bee Catalogue, 2008) = Latin **nāna** f (*genitive* nānae); *first declension* - dwarf (female); [Etym.: from the Latin, *nanus*, dwarf.]*

nebulosa, *Melipona (Michmelia)* Camargo, 1988* (Moure's Bee Catalogue, 2008) = **from Latin nebulōsus** m (*feminine* nebulōsa, *neuter* nebulōsum); *first/second declension* - misty, foggy, clouded

nebulata, *Meliponula (Meliplebeia)* (Smith)^ (Eardley, 2004) = From Engish nebulated - Clouded with indistinct colour markings

nigerrima, *Trigona* Cresson, 1878* (Moure's Bee Catalogue, 2008) = Latin **nigerrimus** (*superlative of* **niger**) - blackest, very black

nigra, *Melipona* Lepeletier, (1836)* (Moure's Bee Catalogue, 2008) = **From Latin niger** m (*feminine* nigra, *neuter* nigrum); *first/second declension* - black

nigrescens, *Melipona (Michmelia)* Friese, 1900* (Moure's Bee Catalogue, 2008) = Present active participle of *nīgrēscō* ("blacken, darken"). –**nīgrēscēns** blackening, darkening, blackish

nigribasis, *"Trigona (Lepidotrigona) ventralis"* forma nov. Sakagami 1975: 71: [Nomen nudum][#] (Rasmussen, 2008) = nigri - New Latin; from *niger* ("black") + **basis** - Latin - An underlying condition or circumstance

nigriceps, *Plebeia* (Friese, 1901)* (Moure's Bee Catalogue, 2008) = nigri - New Latin; from *niger* ("black") + *-ceps* ("headed"). **nigriceps** *m, f, n* (*genitive* **nigricipis**); *third declension* - black-headed

nigrifacies, *Trigona beccarii* var. Friese, 1912^ (Eardley, 2004) = nigri - New Latin; from ***niger*** ("black") + **faciēs** *f* (*genitive* **faciēī**); *fifth declension* – face; having a black face

nigrior, *Partamona* (Cockerell, 1925) and syn. [*Trigona cupira nigrior* Cockerell, (1925)]* (Moure's Bee Catalogue, 2008) = nigri (From Latin **niger** - black) + -or (suffix **–or** *used to form an abstract noun*)

nigripes, *Plebeia* (Friese, 1901)* (Moure's Bee Catalogue, 2008) = nigri (From Latin **niger** - black) + pes (Latin – "foot"); having black feet

nigrilabris, *Partamona* Pedro & Camargo, (2003) [*Etym.*: from the Latin, *niger* = black, + *labrum* = lip, referring to the black colour of the labrum..]* (Moure's Bee Catalogue, 2008)

nigrita, *Trigona pallida* subsp. Friese, (1901)* (Moure's Bee Catalogue, 2008) = Latin **nīgritus** m (*feminine* nigrīta, *neuter* nigrītum); *first/second declension* - black

nigritula, *Trigona* Müller, F. (1921)* (Moure's Bee Catalogue, 2008) = From Latin nigritus - black + ula (Latin suffix –ula –"diminutive form") see also *nigrita*

nigrohirta, *Nannotrigona (Scaptotrigona)* Posey,. (1983)* (Moure's Bee Catalogue, 2008) = nigro (From Latin **niger** - black) + hirta (Latin **hirta – "hairs"**); having black hairs

nitidifrons, *Melipona (Michmelia)* Benoist, 1933* (Moure's Bee Catalogue, 2008) = nitidi - Latin *nitidus* ("shining") + frons (see **frons**); having a shining forehead

nitidiventris, *Lepidotrigona* (Smith, 1857)[#] (Rasmussen, 2008) = nitidi - Latin *nitidus* ("shining") + ventris (Latin – "abdomen"); having a shiny abdomen. Sp. 62

Sp. 62 Lepidotrigona *nitidiventris Photo: Zesrin Soh*

nitidula, *Trigona* Cockerell, (1917)* (Moure's Bee Catalogue, 2008) = From Latin *nitidus* ("shining") + *-ulus*. –- somewhat elegant or trim

obscura, *Oxytrigona* (Friese, 1900) and subsp. *Trigona (Oxytrigona) obscura daemoniaca* Nates Parra, (1983)* (Moure's Bee Catalogue, 2008) = From Latin **obscūrus** m (*feminine* obscūra, *neuter* obscūrum); *first/second declension* - dark, dusky; see also ***obscurior***

obscuripes, *Trigona fulvipes* subsp. Friese, H. (1916)* (Moure's Bee Catalogue, 2008) = From Latin *obscurus* ("dark") + *pēs* ("foot"). - (*genitive* obscuripedis); *third declension* - dark-footed

obscurior, *Melipona (Eomelipona)* Moure, 1971* (Moure's Bee Catalogue, 2008) **= Latin - obscūrior** (*comparative of* obscūrus) - darker, duskier

ochrotricha, *Scaptotrigona* (Buysson, in Du Buysson & Marshall, 1892)* (Moure's Bee Catalogue, 2008) = **ochro** from Ancient Greek ὤχρα (*ōkhra*, "pale yellow") + **tricha-** From Ancient Greek *θρίξ* (thríks, "hair"); having pale yellow hairs.

ochrotrichus, *Trigona* Du Buysson. & Marshall, (1892) syn. to *Scaptotrigona* ***ochrotricha*** (Buysson, in Du Buysson & Marshall, 1892)* (Moure's Bee Catalogue, 2008), see ***ochrotricha***

opaca, *Paratrigona* (Cockerell, 1917)* (Moure's Bee Catalogue, 2008) **= From Latin opācus** m (*feminine* opāca, *neuter* opācum); *first/second declension* - shaded, darkened

ornata, *Austroplebeia* (Rayment, 1932)[#] (Rasmussen, 2008)= From Latin **ōrnātus** m (*feminine* ōrnāta, *neuter* ōrnātum); *first/second declension* - adorned, decorated, having been adorned.

ornaticeps, *Paratrigona* (Schwarz, 1938) and variety *Paratrigona (Paratrigona) ornaticeps guigliae* Moure, J. S. (1960)* (Moure's Bee Catalogue, 2008) = ornate (Latin - "decorated") + ceps (Latin - "head"); having decorated head

pachysoma, *Trigona* Cockerell, (1917)* (Moure's Bee Catalogue, 2008) = From Ancient Greek *παχύς* (pakhus, "thick") - "Thick, or causing or relating to thickness." + soma from Ancient Greek *σῶμα* (sōma, "body").; having a thick body.

pallens, *Trigona* (Fabricius, 1798)* (Moure's Bee Catalogue, 2008) = Latin **pallēns** m, f, n (*genitive* pallentis); *third declension* - blanching, pale,

pallida, *Trigona (Trigona)* var. *muzoensis* Schwarz, H. F. (1948)* (Moure's Bee Catalogue, 2008) = Latin - **pallidus** m (*feminine* pallida, *neuter* pallidum); *first/second declension* - pale, pallid

pallidicincta, *Trigona* Cockerell 1918[#] (Rasmussen, 2008) = Latin - **pallidus** m (*feminine* pallida, *neuter* pallidum); *first/second declension* - pale, pallid + cincta (Latin "girded or bordered"); very clearly girded

pallidistigma, *Trigona pallidistigma* **Cameron** 1908 syn to ***Tetragonula fuscobalteata*** (Cameron, 1908)[#] (Rasmussen, 2008) = Latin - **pallidus** m (*feminine* pallida, *neuter* pallidum); *first/second declension* - pale, pallid + Latin -(*genitive* stigmatis); n, *third declension* - brand (burned mark,); having a pale scar or burned mark.

pannosa, *Paratrigona* Moure, 1989* (Moure's Bee Catalogue, 2008) = From Italian **pannoso** m (*feminine* pannosa, *masculine plural* pannosi, *feminine plural* pannose) - creamy

paranigra, *Frieseomelitta* (Schwarz, 1940) and subsp. *Trigona (Tetragona) nigra* ***paranigra*** Schwarz, H. F. (1940)* (Moure's Bee Catalogue, 2008) = Latin **parā** - ("produce, procure, bring forward, bring forth"), + nigra (Latin - black)

parvula, *Liotrigona* Darchen, 1971a^ (Eardley, 2004) = Latin **parvulus** m (*feminine* parvula, *neuter* parvulum); *first/second declension* - little, petty, slight.

pectoralis*, Scaptotrigona* (Dalla Torre, 1896)* (Moure's Bee Catalogue, 2008) = Latin **pectorālis** m, f (*neuter* pectorāle); *third declension* - of or pertaining to the chest or breast. Sp. 63

Sp. 63 *Scaptotrigona pectoralis* Photo: Melipotrgona De Costa Rica Melipona

pediculana, Trigonisca (Fabricius, 1804)* (Moure's Bee Catalogue, 2008) = Latin **pedīculus** m (*genitive* pedīculī); *second declension* - A little foot. + ana (Latin – "dimunitive form")

peltata*, Paratrigona* (Spinola, 1853)* (Moure's Bee Catalogue, 2008) = From Latin *pelta* ("a small crescent-shaped shield"). **peltātus** m (*feminine* peltāta, *neuter* peltātum); *first/second declension* - armed with a pelta; shield-bearing *Sp. 64*

Sp. 64 Paratrigona peltata Photo: Melipotrigonas De Costa Rica

pellucida*, Trigona* Cockerell, 1912* (Moure's Bee Catalogue, 2008) = From Latin **pellūcidus** m (*feminine* pellūcida, *neuter* pellūcidum); *first/second declension* - transparent, pellucid, very bright

penna*, Hypotrigona* sp. n. Eardley 2004, [Etym.: The name describes the plumose vestiture on the vertex and scutum. Distribution: This species is confined to West Africa.]^ (Eardley, 2004) = From earlier Latin **petna* ("feather, wing"), - Modern **Latin penna** f (*genitive* pennae); *first declension* - feather, fin, wing

permixta*, Paratrigona* Camargo & Moure, 1994 [*Etym.:* from the Latin: *permixtus* = indicating the intermediary condition of the species, which combinates characters from different groups.]* (Moure's Bee Catalogue, 2008)

perangulata*, Tetragona* (Cockerell, 1917)* (Moure's Bee Catalogue, 2008) = per (Latin prefix per- -"very") + angulata (Latin for "of angular" attributes).; being very angular

percincta*, Austroplebeia* (Cockerell, 1929)[#] (Rasmussen, 2008) = per (Latin prefix per- - "very") + cincta (Latin "girded or bordered"); very clearly girded. see also *cincta*

perilampoides, *Nannotrigona* (Cresson, 1878)* (Moure's Bee Catalogue, 2008) = peri (Latin "bee larvae") + lamp (Latin - "light") + oides (Latin suffix **-oides** "resembling") – bee larvae resembling lights. Sp. 65

Sp. 65 *Nannotrigona perilampoides* Photo Luis Guillermo Vera

pernigra, *Melipona (Michmelia) seminigra* subsp. Moure & Kerr, 1950* (Moure's Bee Catalogue, 2008) = From Latin Prefix **'per-'** "very" something. + nigra (Latin – "black"); being very black

pigra, *Trigona* Müller, F. (1875). [*Etym.:* from the Latin, *piger, pigritiosus*, lazy, slow, sluggish.]* (Moure's Bee Catalogue, 2008)

pilicornis, Camargoia (Ducke, 1910)* (Moure's Bee Catalogue, 2008) = pili (Latin –"A hair") + cornis (Latin – horn, antenna): having antennae with hairs.

planifrons, *Platytrigona* (Smith, 1865)[#] (Rasmussen, 2008) = From **plānus** *first/second declension* - level, flat, even + frons (Latin – "forehead"); having a flat forehead

polysticta, *Scaptotrigona* Moure, 1950* (Moure's Bee Catalogue, 2008) = from Greek poly-, combining form of polys "much" (plural polloi) + sticta from Greek στικτα stikta, 'spotty'; having many spots.

postica, ***Scaptotrigona*** (Latreille, 1807)* (Moure's Bee Catalogue, 2008) = From Latin **postīcus** m (*feminine* postīca, *neuter* postīcum); *first/second declension* - back, rear, posterior. Sp. 80

Sp. 66 *Scatotrigona postica* Photo: Giorgio Venturieri

psilosoma, *Trigonisca* Albuquerque, P. M. C. (1992) (Nomen nudum)* (Moure's Bee Catalogue, 2008) = psilo from Greek ψιλο filo, "thin" + soma From New Latin, from Ancient Greek *σῶμα* (sōma, "body").; having a thin body.

pugnax, *Plebeia* Françoso Jr.,. & Bicudo, (2000)* (Moure's Bee Catalogue, 2008) = From Latin *pugnō* ("I fight, -combat"). **pugnāx** m, f, n (*genitive* pugnācis); *third declension* - combative, fond of fighting

pulchella, *Melipona* Müller, F. (1875)* (Moure's Bee Catalogue, 2008) = **From Latin pulchellus** m (*feminine* pulchella, *neuter* pulchellum); *first/second declension* - beautiful little

pulchra, *Plebeia* Ayala, 1999 [Etym.: from the Latin, *pulchrum*: = pretty.] * (Moure's Bee Catalogue, 2008)

punctata, *Nannotrigona* (Smith, 1854)* (Moure's Bee Catalogue, 2008) = From Latin **punctātus** m (*feminine* punctāta, *neuter* punctātum); *first/second declension* - punctuated, pointed

puncticollis, *Melipona (Eomelipona)* Friese, 1902* (Moure's Bee Catalogue, 2008) = New Latin; from *punctus* ("pierced") + *collum* ("stem"). - having a pierced stem

pusilla, *Leurotrigona* Moure, et al. (1988)* (Moure's Bee Catalogue, 2008) = From Latin **pusillus** m (*feminine* pusilla, *neuter* pusillum); *first/second declension* - very little, very small This is the smallest known meliponine (body length ca. 2.0 mm). The nest is a cavity of 4 cm in length. (Camargo, 2008)

pura, *Trigona* Cockerell, T. D. A (1920)* (Moure's Bee Catalogue, 2008) = From Latin **pūrus** m (*feminine* pūra, *neuter* pūrum); *first/second declension* - clear, limpid, clean; pure

quadrangular, *Tetragona* (Lepeletier, 1836)* (Moure's Bee Catalogue, 2008) = From Latin **quādrangulus** *first/second declension* - quadrangular (four-sided)

quadrifasciata, *Melipona (Melipona)* Lepeletier, 1836 * (Moure's Bee Catalogue, 2008) = quad- (Latin prefix meaning "four".) + Latin fasciata – "banded"; having four bands or four banded Sp. 67

quadripunctata, *Schwarziana* (Lepeletier, 1836)* (Moure's Bee Catalogue, 2008) = quadric (Latin – square or regular form) + punctata (From Latin **punctātus** - punctuated, punctured)

quinquefasciata, *Melipona (Melikerria)* Lepeletier, 1836* (Moure's Bee Catalogue, 2008) = Latin quinque – "five" + Latin fasciata – "banded"; having five bands or five banded

Sp. 67 Melipona quadrifasciata Photo: Luia Carlos Da Cunha

recursa, *Trigona* Smith, 1863* (Moure's Bee Catalogue, 2008) = English **recursant** (*not comparable*) - (heraldry) Having its back towards the observer;

rhodoptera, *Trigona* Cockerell, (1912)* (Moure's Bee Catalogue, 2008) = From Ancient Greek ῥόδον (rhodon, "rose"). + ptera (Latin – "wings").; Having rose-coloured wings.

robustior, *Trigona (Trigona) hypogea* var. Schwarz (1940)* (Moure's Bee Catalogue, 2008) = Latin **robustior** (*comparative of* robustus) - harder, firmer, more robust

rubra, *Melipona marginata* Schwarz, (1932)* (Moure's Bee Catalogue, 2008) = From Latin **ruber** *first/second declension* - red (colour), ruddy; (New Latin) Used as a species epithet.

rufibasalis, *Tetragonilla* (Cockerell, 1918)[#] (Rasmussen, 2008) = rufi (Latin – "reddish") + basalis - Latin -m *and* f, **basale** n; *third declension* - basal, having a pedestal or foot.

ruficornis, *Tetragonula* (Smith in Horne & Smith, 1870)[#] (Rasmussen, 2008) = rufi (Latin – "reddish") + cornis ("horn, antenna"); having reddish antennae. Sp. 68

Sp. 68 *Tetragonula ruficornis* Photo: Claus Rasmussen

rufescens, *Melipona (Michmelia)* Friese, 1900* (Moure's Bee Catalogue, 2008) = **Latin rūfēscēns** m, f, n (*genitive* rūfēscentis); *third declension* - reddening, reddish

ruficrus, *Melipona* Ducke, 1916* (Moure's Bee Catalogue, 2008) = rufi (Latin – "reddish") + crus (Latin – "legs") having reddish-brown legs

rufipes, *Lestrimelitta* (Friese, 1903) and *Trigona (Lestrimelitta) limao rufipes* Friese, (1903)* (Moure's Bee Catalogue, 2008) = rufi (Latin – "reddish") + pes (Latin – "foot"); having reddish feet

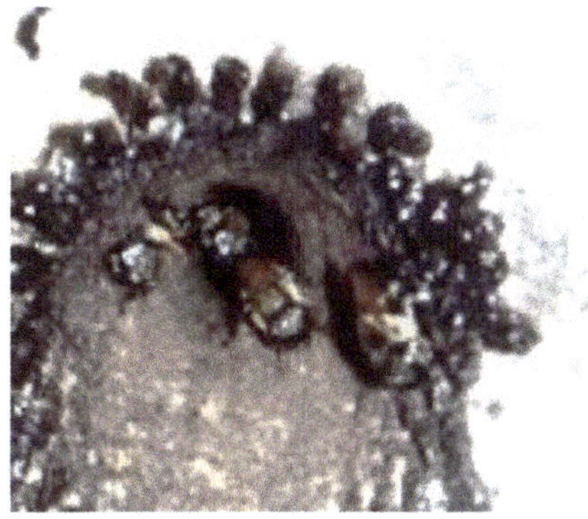

Sp. 69 *Melipona rufiventris* Photo: Wilson Gussoni

rufiventris, *Melipona (Michmelia)* Lepeletier, 1836* (Moure's Bee Catalogue, 2008) = rufi (Latin – "reddish") + ventris (Latin – "abdomen"); having a red abdomen. Sp. 69

rufoides, *Trigona* Strand, 1910* (Moure's Bee Catalogue, 2008) = From Latin **rūfus** - red, ruddy + oides (Latin - resemble).; resembling ruddy coloured.

scapisetosa, *Paratrigona* Gonzalez & Griswold, 2011 [*Etym.:* the specific epithet is a combination of the Latin words *scapus* and *setae*, in reference to the distinctly long hairs along the inner margin of the antennal scape.]* (Moure's Bee Catalogue, 2008)

scintillans, *Trigona* Cockerell 1920a syn. to *Lisotrigona cacciae* (Nurse, 1907)[#] (Rasmussen, 2008) = Latin - **scintillāns** m, f, n (*genitive* scintillantis); *third declension* - sparkling **Sp. 5**

scutella, *Melipona* Lepeletier, A. L. M. (1836)* (Moure's Bee Catalogue, 2008). = English **scutella** (*plural* scutellae) - A scutellum. In the shield-shaped structures in insects (zoology), the **scutellum** is the posterior portion of either the mesonotum or the insect's thorax metanotum. It follows the scutum and is followed by the small postscutellum. (Wikipedia, Scutellum, 2013)

Sp. 70 *Melipona seminigra* Photo: Maria Rosa bernasconi

scutellaris, *Melipona (Michmelia)* Latreille, 1811* (Moure's Bee Catalogue, 2008) = Latin **scutellaris** m, f (*neuter* scutellare); *third declension* - saucer-shaped see also ***scutella.***

seminigra, *Melipona (Michmelia)* Friese, 1903* (Moure's Bee Catalogue, 2008) = semi (Latin – "partial"-) + nigra(Latin – "black"); partially black Sp. 70

sericea, *Trigona* **Friese** 1933a syn. to *Tetrigona apicalis* (Smith, 1857)[#] (Rasmussen, 2008) = **Latin - sēriceus** *first/second declension* - Made of silk, silken.

Sp. 72 *Melipona subnitida* Photo via Maria Rosa Bernasconi

silacea, *Proplebeia* (Wille, 1959) [*Etym.*: from the Latin, *silaceous* = ocher.]* (Moure's Bee Catalogue, 2008)

spinosa, *Lestrimelitta* Marchi & Melo, 2006 [Etym.: from the Latin, *spinosus*, with thick setae on the pronotal lobes, tegulae and mesepisterna.]* (Moure's Bee Catalogue, 2008) = Latin **spīnōsus** - thorny, prickly;

Sp. 71 *Tetrigona melanoleuca* Photo: Ahmad Zaini

spinipes, *Trigona* (Fabricius, 1793)* (Moure's Bee Catalogue, 2008) = From Latin *spineus* ("spiny") + *pēs* ("foot"). – **spinipēs** m, f, n (*genitive* spinipedis); *third declension* - spiny-footed, spiny-legged. Sp. 31

stigma, *Trigona* Smith, F. (1854)* (Moure's Bee Catalogue, 2008) = Latin -(*genitive* stigmatis); n, *third declension* - brand (burned mark,)

subfulva, *Geotrigona* Camargo & Moure, 1996* (Moure's Bee Catalogue, 2008) = **sub-** (Latin prefix - "subsidiary") + from Latin **fulvus** - reddish-yellow, tawny, amber-coloured, fulvous; somewhat yellowish

subgrisea, *Geotrigona* (Cockerell, 1920)* (Moure's Bee Catalogue, 2008) = **sub-** (Latin prefix - "subsidiary") + Latin griseus - (New Latin) grey ; somewhat greyish

subnigra, *Geotrigona* Schwarz, 1940* (Moure's Bee Catalogue, 2008) = From Latin **subniger** m (*feminine* subnigra, *neuter* subnigrum); *first/second declension* -blackish, somewhat black

subnitida, *Melipona (Melipona)* Ducke, 1910* (Moure's Bee Catalogue, 2008) = **sub-** (Latin prefix - "subsidiary") + nitida - Latin *nitidus* ("shining"); partially shining. Sp. 72

subnuda, *Paratrigona* Moure, 1947 and alt. comb. *Paratrigona lineata* ***subnuda*** Moure, (1947)* (Moure's Bee Catalogue, 2008) = **sub-** (Latin prefix - "subsidiary") + nuda (Latin – "naked"); partially naked.Sp. 73

Sp. 73 *Paratrigona subnuda* Photo: Luia Carlos Da Cunha

subobscuripennis, *Scaptotrigona* (Schwarz, 1951) and alt. comb. *Trigona (Scaptotrigona) mexicana* ***subobscuripennis*** Schwarz, (1951)* (Moure's Bee Catalogue, 2008) = **sub-** (Latin prefix - "subsidiary") + obscuri (Latin – "obscure) + pennis (Latin – "wings"); having partially obscured wings.

subtilis, *Partamona* Pedro & Camargo, 2003 [*Etym:* from the Latin, *subtilis*, slim, slender.]* (Moure's Bee Catalogue, 2008)

tantilla, *Proplebeia* Camargo, Grimaldi & Pedro, 2000 [*Etym.:* from the Latin, *tantillus*, little, small.]* (Moure's Bee Catalogue, 2008)

tenuis, *Scaura* (Ducke, 1916) also syn. *Melipona crassipes* ***tenuis*** Ducke, A. (1916)* (Moure's Bee Catalogue, 2008) = Latin **tenuis** - thin, fine, slender, delicate, subtle

terminata, *Lepidotrigona* (Smith, 1878)# (Rasmussen, 2008) = Latin participle derived from **terminate** - Terminated; limited; bounded; ended., Having a definite and clear limit or boundary. Sp. 74

Sp. 74 *Lepidotrigona terminata* Photo: Azmi Bin Awang Azmi

testacea, *Trigona* Spinola, M. (1853) and also *Partamona* ***testacea*** (Klug, 1807)* (Moure's Bee Catalogue, 2008) = from English **testaceous** - Of a reddish-yellow or brownish colour, like a brick.

testaceicornis, *Nannotrigona* (Lepeletier, 1836)* (Moure's Bee Catalogue, 2008) = **testacei** (from

Sp. 75 Nannotrigona testaceicornis Photo: via maria Rosa bernasconi

English **testaceous** - brownish) + cornis ("horn, antenna"); having brownish antennae. Sp. 75

Sp. 76 *Tetragonula testaceitarsis* Photo: Ahmad Zaini

testaceichelis, *Trigona erythra* var. Strand, 1911c^ (Eardley, 2004) = **testacei (**from English **testaceous** - brownish**)** + chelis ("claws"); having brownish claws

testaceinerva, *Trigona* Cameron 1908 *Tetragonula melanocephala* (Gribodo, 1893)[#] (Rasmussen, 2008) = **testacei (**from English **testaceous** - brownish**)** + **nerva** – from nerver - to cord (cover with cords, ropes or strings); having brownish veins.

testaceitarsis, *Tetragonula* (Cameron, 1901)[#] (Rasmussen, 2008) = **testacei (**from English **testaceous** - brownish**)** + tarsis (insect tarsal segment) ; having brownish tarsis. Sp. 76

thoracica, *Geniotrigona* (Smith, 1857)[#] (Rasmussen, 2008) and *Trigona thoracica* Cresson, E. T. (1878)* = From Latin-**thōrācicus** English **thoracic** (*not comparable*) - (anatomy) Of the thorax. Sp. 33

titiana, *Melipona (Michmelia)* (Gribodo, 1893)* (Moure's Bee Catalogue, 2008) = English **titian** (*plural* titians) - A bright auburn colour, tinted with gold. The Italian painter *Titian* (Italian: Tiziano Vecellio) made frequent use of this colour. Sp. 77

tomentosa, *Trigona* Friese, 1900^ (Eardley, 2004) = From Latin **tōmentōsus** m (*feminine* tōmentōsa, *neuter* tōmentōsum); *first/second declension* - having a mass of rough hairs

Sp. 77 *Melipona titiana* Photo: Claus Rasmussen

torrid, *Melipona (Eomelipona)* Friese, 1916* (Moure's Bee Catalogue, 2008) = From Latin *torridus*, from *torrere* ("parch, scorch").

trichocerata, *Frieseomelitta* Moure, 1990* (Moure's Bee Catalogue, 2008) = From **Ancient Greek** ***θρίξ*** (thríks, "hair") + cerata (From Latin **cērātus** - waxed); having waxy hairs

tricolorata, *Scaptotrigona* Camargo, 1988* (Moure's Bee Catalogue, 2008) = Latin **tricolor** m, f, n (*genitive* tricoloris); *third declension* - three-coloured + -ata ("of or pertaining to")

Sp. 78 *Tetragona truncata* Photo: Claus Rasmussen

trigonospila, *Tetragona* Moure, J. S. (2000)* (Moure's Bee Catalogue, 2008) = Latin **trigōnōs** - trigonal; three-angled; having three angles + pila("A mortar to grind or mix ingredients with a pestle")

trochanterica *Lepidotrigona* (Cockerell, 1920)[#] (Rasmussen, 2008) = (anatomy) second segment of insects leg + ica (of or pertaining to); pertains to the trochanter.

trombeta, *Melipona* Müller, H. (1875) and alt. comb. *Trigona trombeta foetida* Müller, F. (1921)* (Moure's Bee Catalogue, 2008) = From Italian **trombetta** f (*plural* trombette) (*Diminutive of*: tromba) - Toy trumpet; most likely nest entrance trumpet shaped.

truncata, *Tetragona* Moure, 1971* (Moure's Bee Catalogue, 2008) = **Latin - truncātus** m (*feminine* truncāta, *neuter* truncātum); *first/second declension* - maimed, mutilated, truncated. Sp. 78

truculenta, *Trigona* Almeida, 1984* (Moure's Bee Catalogue, 2008) = From Latin **truculentus** m (*feminine* truculenta, *neuter* truculentum); *first/second declension* - savage, ferocious

Sp. 79 *Lepidotrigona ventralis* Photo: Debra Smith

unidentata, *Trigonisca* Albuquerque & Camargo, 2007 [*Etym:* from the Latin, *unus* + *dentatus*, referring to the unidentate mandible.]* (Moure's Bee Catalogue, 2008)

variicolor, *Plebeia* (Ducke, 1916)* (Moure's Bee Catalogue, 2008) = From Latin **varius** m (*feminine* varia, *neuter* varium); *first/second declension* - diverse, different, various, variegated + color

variegatifrons, Trigonisca Albuquerque & Camargo, 2007 [*Etym.:* from the Latin, *variegatus* (=variagated colors) + *fronte* (= frons).]* (Moure's Bee Catalogue, 2008); having a variegated forehead

variegatipes, *Melipona (Melipona)* Gribodo, 1893* (Moure's Bee Catalogue, 2008) = variegati (Latin - Streaked or a variety of colour) + pes(Latin – "foot"); having variegated feet

Sp. 80 *Scaptotrigona postica* xanthotricha
Photo: Luiz carlos Da Cunha

ventralis, *Lepidotrigona* (Smith, 1857)# (Rasmussen, 2008) =From Latin ventrālis, pertaining to venter ("belly, abdomen"). Sp. 79

versicolor, *Trigona versicolor* Friese 1909("1908") Syn. to *Homotrigona fimbriata* (Smith, 1857) # (Rasmussen, 2008) = Latin **versicolor** m, f, n (*genitive* versicolōris); *third declension* - particoloured; of various colours

vestita, *Melipona (Trigona)* Spinola, (1840)* (Moure's Bee Catalogue, 2008) = From Latin **vestītus** m (*feminine* vestīta, *neuter* vestītum); *first/second declension* - (rare) clothed, clad, dressed.

vitrifrons, *Trigonisca* Albuquerque & Camargo, 2007 [*Etym.:* from the Latin, *vitreus* (= glass) + *fronte* (=frons).]* (Moure's Bee Catalogue, 2008) = having a glassy forehead

xantina, *Melipona rufiventris* subsp. Mercado, (1968)* (Moure's Bee Catalogue, 2008) = From French xanthine, ultimately from Ancient Greek ξανθός (xanthós), "yellow", because xanthine is yellowish-white.

xanthogastra, *Partamona* Pedro & Camargo, 1997* (Moure's Bee Catalogue, 2008) = From Latin **xanthogaster** m (*feminine* xanthogastra, *neuter* xanthogastrum); *first/second declension* - having a yellow belly

xanthotricha, *Scaptotrigona postica var.* Moure, (1950)* (Moure's Bee Catalogue, 2008) = xantho (Ancient Greek – "yellow") + tricha (Ancient Greek – "hair); having yellow hair. Sp. 80

xanthopleura*, Frieseomelitta* Francke, et al. (2000)* (Moure's Bee Catalogue, 2008) = xantho (Ancient Greek – "yellow") + pleura ("ribs, sides"); having yellow ribs or sides

xanthopoda, *Geotrigona* Camargo & Moure, 1996* (Moure's Bee Catalogue, 2008) = **xantho** (Ancient Greek – "yellow") + **poda (**New Latin suffix derived from Greek roots meaning "foot(ed)") *Used to form taxonomic names of animals having feet or foot-like parts.*

xanthopus, *Melipona* Schwarz, (1932). (Nomen nudum)* (Moure's Bee Catalogue, 2008) = English xantho ("yellow") + Latin pus ('body"); having a yellow body

zonata, *Trigona (Parapartamona) brevipilosa* Schwarz, (1948)* (Moure's Bee Catalogue, 2008) = From Latin **zōnātus** m (*feminine* zōnāta, *neuter* zōnātum); *first/second declension* - belted, girdled

zonulata, *Melipona zonulata* Spinola, (1840). (Nomen nudum)* (Moure's Bee Catalogue, 2008) = zonula - (anatomy) The small belt-like regions + -ata (see suffix **–ata**)

Chapter 8

General Epithets for Meliponine Species Taxa \

All other epithets have been grouped into this General Epthets section. Most of the epithets here denotes the Meliponine behavioural characteristics and appearances. While a few are uncertain and ambiguous, some are extinct (identified in prehistoric fossils).

General Epithets

aemula, *Trigona* Illiger, K. (1806) (Nomen nudum)* (Moure's Bee Catalogue, 2008) = Latin **aemulus** *m* (*feminine* **aemula**, *neuter* **aemulum**); *first/second declension* - striving to equal, rivalling

afinis, *Melipona compressipes* subsp. Posey, (1983). (Nomen nudum)* (Moure's Bee Catalogue, 2008) = Latin translation of afinis means "from the ends"; From Latin **affīnitās** f (*genitive* affīnitātis); *third declension* - relationship, affinity, union, connection

affinis, *Scaptotrigona* Camargo, (1988) (*Nomen nudum* According to Camargo, 1988: 360, an unpublished name by Schwarz)* (Moure's Bee Catalogue, 2008) = see definition of ***afinis***

amaura, *Partamona* Ayala, (1999). [the name "Partamona amaura", was proposed by Ayala,R. 1992, in his master thesis]* (Moure's Bee Catalogue, 2008) = From Latin *amor*, *amorem*.**amaur** - sweetheart, darling

ambigua, *Melipona (Melikerria)* Roubik & Camargo, 2012 [Etym.: from Latin, *ambiguus*, uncertain, doubtful.]* (Moure's Bee Catalogue, 2008)

anamitica, *Homotrigona* (Friese, 1909)# (Rasmussen, 2008) = From Late Latin *mythicus*, from Ancient Greek *μυθικός* (mythikos).**mítico** - mythic, mythical; **anamitica** - Latin "awesome."

Sp. 81 *Lepidotrigoa arcifera* Photo Claus Rasmussen

arcifera, *Lepidotrigona* (Cockerell, 1929)# (Rasmussen, 2008) = **arci** (Latin – arcī (*genitive* **arcis**); *f, third declension* - citadel, stronghold) + **fera** (Latin - (*genitive* **ferae**)-wild animal, beast) Sp. 81

armata, *Trigona* Magretti, 1895^ (Eardley, 2004) = From Latin **armātus** m (*feminine* armāta, *neuter* armātum); *first/second declension* - armoured

atomella, *Trigona* **Cockerell** 1919b syn. of *Tetragonula fuscobalteata* (Cameron, 1908)# (Rasmussen, 2008) a tiny minute species = atom (smallest indivisible particle), + ella(she (subject form)) From Latin *illa* or see suffix **–ella.**Sp. 83

busara, *Trigona* Cockerell 1918 (synonym of *Lophotrigona canifrons* (Smith, 1857)# (Rasmussen, 2008) Source word basara - 'good sense', from Arabic. Sp. 82

Sp. 82 *Lophotrigona canifrons* Photo: azmand Rosland

cagafogo, *Trigona* Müller, H. (1874)* (Moure's Bee Catalogue, 2008) - [Zool.] – Caga-fogo - Portuguese "Shits fire" is the popular name of a kind of formiga (in Portuguese "ant"), also called foot-washing.

callura, *Melipona quadrifasciata* var. Cockerell, (1919) * (Moure's Bee Catalogue, 2008) = From Latin callur meaning "skilled."

cearina, *Melipona (Trigona)* Vachal, 1903 syn. n.^ (Eardley, 2004) = Latin cēra "wax" - Ointment other excipient serves and is prepared with wax, ceresin (a white wax) and liquid paraffin.

concinnula, *Melipona (Eomelipona)* Cockerell, 1919* (Moure's Bee Catalogue, 2008) = From Latin **concinnus** *first/second declension* neat, elegant, pretty, + -ula (Latin suffix – diminutive form)

Sp. 83 *Tetragonula fuscobalteata* Photo Acai Ibrahim

confusella, *Trigona* Cockerell 1919b# (Rasmussen, 2008) = Latin- confus- "compare" + -ella (Latin suffix indicating a diminutive form)

crinita, *Melipona (Michmelia)* Moure & Kerr, 1950* (Moure's Bee Catalogue, 2008) = **From Latin crīnītus** m (*feminine* crīnīta, *neuter* crīnītum); *first/second declension* - hairy, long

criptica, *Partamona* Pedro & Camargo, 2003 [*Etym.:* from the Greek, *kryptós*, hidden, referring to the resemblance with *Partamona helleri* (Friese).]* (Moure's Bee Catalogue, 2008) = cryptic - From Late Latin *crypticus*, - Having hidden meaning, Mystified or of an obscure nature.

daemoniaca, *Oxytrigona* Camargo, 1984 and alt. comb. *Oxytrigona tataira* ***daemoniaca*** Camargo, (1984). also *Trigona (Oxytrigona) obscura* ***daemoniaca*** (Nomen nudum)* (Moure's Bee Catalogue, 2008) = From Latin daemoniacus *first/second declension* - demonic, devilish

decumana, *Trigona* Spinola, M. (1840). (Nomen nudum)* (Moure's Bee Catalogue, 2008) = From Latin **decumānus** m (*feminine* decumāna, *neuter* decumānum); *first/second declension* - (attributive: "of the tenth")

devicta, *Tetragonula* (Cockerell, 1921) Type locality: Myanmar "bead of clear pale Burmese amber from the Hukong Valley, received from Mr. R.C.J. Swinhoe" (Rasmussen, 2008) = devicta Latin - ablative feminine singular of *dēvictus* - dēvictus m (*feminine* dēvicta, *neuter* dēvictum); *first/second declension* - defeated, subdued. See also ***devictus***

devictus, *Meliponorytes* Cockerell 1921 and new comb. *Trigona (Tetragonula) devicta*; Kerr & Maule, 1964 (redescr. comp. fossil, Miocene) && (Moure J.S. & Camargo J.M.F., 1978). *Meliponorytes devictus* was re-studied and found to be a valid Trigona, close to *T. iridipennis* of the subgenus *(Tetragonula)* and renamed *Trigona (Tetragonula) devicta* (Cockerell) (Kerr W, Maule V., 1964) Sp. 84

Sp. 84 ***Tetragonula iridipennis*** Photo: Claus Rasmussen

diluta, *Melipona* Illiger, K. (1806) (Nomen nudum)* (Moure's Bee Catalogue, 2008) = *From Latin dīluō* ("wash away; dissolve, dilute"). **dīlūtus** *m* (*feminine* **dīlūta**, *neuter* **dīlūtum**); *first/second declension* washed away, drenched, diluted.

dimidiata, *Trigona* Smith, 1854* (Moure's Bee Catalogue, 2008) = From Latin **dīmidiātus** m (*feminine* dīmidiāta, *neuter* dīmidiātum); *first/second declension* - halved (divided in half)

dispar, *Frieseomelitta* (Moure, 1950)* (Moure's Bee Catalogue, 2008) = Latin - **dispar** m, f, n (*genitive* disparis); *third declension* - unlike, dissimilar, different, unequal

dubia, *Melipona (Michmelia)* Moure & Kerr, 1950* (Moure's Bee Catalogue, 2008) = From Latin Shortened form of *duhibius* ("held as two"), from *duo* ("two") + *habeō* ("have, hold"). **dubius** m (*feminine* dubia, *neuter* dubium); *first/second declension* - Moving in two directions alternately, vibrating to and fro, fluctuating, wavering.

eocenica, *Hypotrigona* Kelner-Pillault, 1969a monobasic and original designation, isotypic with *Kelneriapis* Sakagami, 1978. = extinct stingless bee of the Eocene period. [&^] (Kelner-Pillault, 1969, 1970)

extranea, *Trichotrigona* Camargo & Moure, 1983* (Moure's Bee Catalogue, 2008) = **From Latin extrāneus** m (*feminine* extrānea, *neuter* extrāneum); *first/second declension* - foreign, strange (This is the only meliponine that strangely does not build storage pots and does not store any kind of food in its nest;)

extrema, *Trigonisca* Albuquerque & Camargo, 2007 [*Etym.:* from the Latin, *extremus* = last.]* (Moure's Bee Catalogue, 2008)

faceta, *Frieseomelitta silvestrii* Absy, et al. (1984)* (Moure's Bee Catalogue, 2008) = From Latin **facētus** m (*feminine* facēta, *neuter* facētum); *first/second declension* -elegant, fine

fallax, *Melipona (Michmelia)* Camargo & Pedro, 2008 [*Etym.:* From the Latin, fallax = deceitful, deceptive, to leading someone on.]* (Moure's Bee Catalogue, 2008)

faecivora, *Meliplebeia* (Strand 1911)* (Eardley & Urban, 2010) = faeci – the singular of **faex** *f* (*genitive* **faecis**); *third declension* - sediment, dregs + **vora** – From Latin vorare ("devour, eat greedily")

filosofiai, *Trigona (Scaptotrigona)* Kerr, et al. (1963)* (Moure's Bee Catalogue, 2008) = From **Ancient Greek** *φιλοσοφία* (philosophia),- *φίλος* (philos, "beloved") + *σοφία* (sophia, "wisdom"). philosophy

fraterna, *Plebeia* Laroca & Rodriguez-Parilli, 2009* (Moure's Bee Catalogue, 2008) = **fraterna** - From Latin frater "brother" - **frāternus** *m* (*feminine* **frāterna**, *neuter* **frāternum**); *first/second declension* - of or pertaining to a kinsman

geneoscantha, *Trigona (Frieseomelitta) varia* subsp. Akahira & Beig, (1967). (Nomen nudum)* (Moure's Bee Catalogue, 2008) = geneos ("class, race") + cantha (from Latin canthus - "ring")

incerta, Paratrigona Camargo & Moure, 1994 [*Etym.:* from the Latin, *incertus*, referring to the doubt about the taxonomic status conferred by the authors.]* (Moure's Bee Catalogue, 2008)

indecisa, *Melipona (Michmelia)* Cockerell, 1919* (Moure's Bee Catalogue, 2008) = Italian & Spanish **indeciso** - indecisive, uncertain, undecided, unresolved

Sp. 85 *Melipona interrupta* Photo: Marcos Ferreira

integrior, *Melipona anthidioides* Cockerell, (1919)* (Moure's Bee Catalogue, 2008) = Latin **integrior** (*comparative of* integer) - more complete, intact

intermedia, *Plebeia* (Wille, 1960)* (Moure's Bee Catalogue, 2008) = Latin **intermedius** m (*feminine* intermedia, *neuter* intermedium); *first/second declension* – "intermediate; which is between"

intermixta, *Melipona* Cockerell, (1918)* (Moure's Bee Catalogue, 2008) = Latin **inter** - between, among + mixta(From Latin **mixtus** - "**mixed,** having been mixed")

interrupta, *Melipona (Melikerria)* Latreille, 1811* (Moure's Bee Catalogue, 2008) = **From Latin interruptus** m (*feminine* interrupta, *neuter* interruptum); *first/second declension* - broken apart, interrupted.Sp. 85

inusitata, *Geotrigona* Moure,. & Camargo (1992)* (Moure's Bee Catalogue, 2008) = From Latin - **inūsitātus** m (*feminine* inūsitāta, *neuter* inūsitātum); *first/second declension* - unusual, extraordinary

iris, *Melipona* Lepeletier, A. L. M. (1836).*Incertae sedis** (Moure's Bee Catalogue, 2008) = Latin **iris** (plural of ira) **īra** *f* (*genitive* īrae); *first declension* - ire, anger, wrath

laboriosa, *Trigona* Smith, (1862)* (Moure's Bee Catalogue, 2008) = From Latin **labōriōsus** m (*feminine* labōriōsa, *neuter* labōriōsum); *first/second declension* - industrious, hard-working

languid, *Frieseomelitta* Moure, 1990* (Moure's Bee Catalogue, 2008) = From Latin **lánguido** m (*feminine* lánguida, *masculine plural* lánguidos, *feminine plural* lánguidas) - languid, weak

lateralis, *Melipona (Michmelia)* Erichson, 1848* (Moure's Bee Catalogue, 2008) = From Latin - **laterālis** m, f (*neuter* laterāle); *third declension* - lateral; of or belonging to the side

limbata, *Trigona* Dominique, J. (1898). (*Nomen nudum*) *Incertae sedis** (Moure's Bee Catalogue, 2008) = From Latin .**limbātus** *m* (*feminine* limbāta, *neuter* limbātum); *first/second declension* - edged, bordered

littoralis, *Partamona* Pedro & Camargo, 2003 [*Etym.:* from the Latin, *littorale* = coastal zone, where the species occurs.]* (Moure's Bee Catalogue, 2008) Usage notes*:* Specifically refers to the *water* at the shore, rather than the land, particularly in the phrase *littoral zone*.

maesta, *Melipona* Puls, J. C. (1868) (Nomen nudum) Incertae sedis* (Moure's Bee Catalogue, 2008) = **maestus** *m* (*feminine* maesta, *neuter* maestum); *first/second declension* - melancholy, gloomy

mirandula, *Nogueirapis* (Cockerell, 1917)* (Moure's Bee Catalogue, 2008) = From Latin **mīrandus** m (*feminine* mīranda, ***neuter*** mīrandum); *first/second declension* - which is to be wondered or marvelled at + -ula(The suffix **–ula**(*-ulus)* is added to a noun to form a diminutive of that noun.)

modesta, *Frieseomelitta* Posey, (1983) (Nomen nudum)* (Moure's Bee Catalogue, 2008) = From Latin **modestus** m (*feminine* modesta, ***neuter*** modestum); *first/second declension* - calm, restrained, modest, reserved,

molesta, *Plebeia* (Puls, in Strobel, 1868)* (Moure's Bee Catalogue, 2008) = Latin - **molestus** m (*feminine* molesta, ***neuter*** molestum); *first/second declension* - troublesome, annoying,

mutata, *Melipona* Lepeletier, (1836)* (Moure's Bee Catalogue, 2008) = From Latin **mūtātus** m (*feminine* mūtāta, ***neuter*** mūtātum); *first/second declension* - mutated, varied, altered,

necrophaga, *Trigona* Camargo & Roubik, 1991* (Moure's Bee Catalogue, 2008) = Composed by *necro-* + *-phage*, from Ancient Greek *νεκροφάγος* (nekrophágos), from *νεκρός* (necros, "cadaver") and *φάγος* (phagos, "one who eats"). From English **necrophage** (*plural* necrophages) - An organism that eats dead or decaying flesh.

nuda, *Paratrigona* (Schwarz, 1943)* (Moure's Bee Catalogue, 2008) = From Latin **nuda** Feminine of *nudus* - **nuda** f - naked

oblitescens, *Melipona interrupta* subsp. Cockerell, (1919)* (Moure's Bee Catalogue, 2008) = From Latin *oblitus*, perfect passive participle of *oblinō* ("smear over"). erase, efface, blot out.+ scens(see **scens**)

opposita, *Melipona* Gribodo, (1893) (junior subjective synonym)* (Moure's Bee Catalogue, 2008) = From Latin **oppositus** m (*feminine* opposita, ***neuter*** oppositum); *first/second declension* - opposite; against

paupera, *Frieseomelitta* (Provancher, 1888)* (Moure's Bee Catalogue, 2008) = Latin paupera - nominative neuter plural of ***pauper*** - m, f, n (*genitive* pauperis); *third declension* - Poor.

permodica, *Trigona* Almeida, 1995* (Moure's Bee Catalogue, 2008) = From *per-* + *modicus* – *Latin* **permodicus** *first/second declension* - very moderate, temperate or restrained

Sp. *86 Plebeia remota* Photo: via Maria Rosa Bernasconi

praeterita, *Tetragonula* (Walker, 1860)[#] (Rasmussen, 2008) = Latin - **praeteritus** m (*feminine* **praeterita**, *neuter* **praeteritum**); *first/second declension* - disregarded, neglected, omitted Sp. 87

Sp. 87 *Tetragonula praeterita* Photo: Claus Rasmussen

pseudocentris, *Melipona* Cockerell, (1912)* (Moure's Bee Catalogue, 2008) = pseudo (Latin – "false") + centris (The genus of large apid bees)

remota, *Plebeia* (Holmberg, 1903)* (Moure's Bee Catalogue, 2008) = From Latin **remōtus** m (*feminine* remōta, *neuter* remōtum); *first/second declension* - remote, distant, far off Sp. *86*

sapiens, *Tetragonula* (Cockerell, 1911)[#] (Rasmussen, 2008) = Latin **sapiēns** m, f, n (*genitive* sapientis); *third declension* - discerning, wise, prudent, discreet Sp. 88

Sp. 88 *Tetragonula sapiens* Photo: Suardi Saide

sesquipedalis, *Trigona*, Almeida, 1984* (Moure's Bee Catalogue, 2008) = Latin *sesqui-* ("one and a half") + *pedal* ("foot, of the foot"). - **sesquipedal** - Measuring or containing a foot and a half.

sevocans, *Trigona* Cockerell, (1922)* (Moure's Bee Catalogue, 2008) = Latin **sēvocāns** m, f, n (*genitive* sēvocantis); *third declension* - separating

sexpunctata, *Melipona* Lepeletier, (1836)* (Moure's Bee Catalogue, 2008) = sex (Latin – "six") + punctata (Latin – "points or punctures"); having six points.

sicophanta, *Melipona* Gribodo, (1893)* (Moure's Bee Catalogue, 2008) = From Latin **sȳcophanta** m (*genitive* sȳcophantae); *first declension* - A snitch, informant

similis, *Lestrimelitta* Marchi & Melo, 2006 [Etym.: from the Latin, *similis*, similar.]* (Moure's Bee Catalogue, 2008)

sóror, *Melipona* Dominique, (1898)* (Moure's Bee Catalogue, 2008) = Latin **soror** f (*genitive* sorōris); *third declension* - sister, female friend

spaca, *Trigona* Müller, (1875)* (Moure's Bee Catalogue, 2008) = Esperanto **spaca** (*plural* spacaj, *accusative singular* spacan, *accusative plural* spacajn)spatial (of or relating to space)

squamuligera, *Melipona* Benoist 1937 (Eardley & Urban, 2010) = squamule English **squamule** (*plural* **squamules**) - A small **scale**. + gera From Latin *generāre*. to generate; [Note: The publication of Benoist 1937 was omitted from the 2004 revision of Eardley. This well-characterized by its hairy species is easily identified by its original description. Type Benoist examined.] (A. Pauly & N.Vereecken, 2013)

succini, *Meliponorytes* Tosi = From Latin sucinum meaning "amber" The fossil is considered close enough to *Melipona bocandei*(Spinola) to justify joining them in one genus. (Kerr W, Maule V., 1964)

suffragata, *Trigona* Cockerell, (1922)* (Moure's Bee Catalogue, 2008) = From Latin **suffrāgātus** m (*feminine* suffrāgāta, *neuter* suffrāgātum); *first/second declension* - supported

timida, *Schwarzula* (Silvestri, 1902)* (Moure's Bee Catalogue, 2008) = From Latin **timidus**(From *timeō* ("I fear") + *-idus*.) m *first/second declension* - fearful, afraid, apprehensive, timid

tristella, *Nannotrigona* Cockerell, 1922* (Moure's Bee Catalogue, 2008) = From Latin **trīstis** m, f (*neuter* trīste); *third declension* - (of smell) foul, offensive + ella (form a diminutive)

turusiri, *Scaptotrigona* (Janvier, 1955) [*Typ.loc:* Yanakachi, Bolivia]* (Moure's Bee Catalogue, 2008) = **turusiri** - Bolivian folk name of the species.

varia, *Tetragona* subsp. *dispar* Moure, (1950)* (Moure's Bee Catalogue, 2008) = From Latin – *varius* ("diverse, various, variegated").

variabilis, *Melipona* Müller, H. (1875)* (Moure's Bee Catalogue, 2008) = Latin **variābilis** *m, f* (*neuter* **variābile**); *third declension* variable, changeable

varians, *Trigona* Müller, F. (1921)* (Moure's Bee Catalogue, 2008) = From Old French **variance** (*plural* **variances**) The act of varying or the state of being variable; The state of differing or being in conflict

vetula, *Liotrigona* Moure & Camargo[&&] (Moure J.S. & Camargo J.M.F., 1978), = **Vetula** means "old woman" in Latin This is the first meliponine bee discovered in copal but of unknown geologic age or geographic origin.

vetusta, *Proplebeia* Camargo, Grimaldi & Pedro, 2000 [*Etym.:* from the Latin, *vetustus*, old, ancient.]* (Moure's Bee Catalogue, 2008)

vicina, *Partamona* Camargo, 1980 and *Melipona vicina* Lepeletier, (1836)* = Latin **vīcīna** f (*genitive* vīcīnae); *first declension* - (female) neighbour/neighbour This bee, as the first line of defence, builds a labyrinth after its entrance as a false nest through which an atrium and then a tunnel leads to the true nest. (A. R. Rech & M. L. Absy, 2011)

vittigera, *Parapartamona* Moure, 1995* (Moure's Bee Catalogue, 2008) = From Latin **vitta** f (*genitive* vittae); *first declension* - band, ribbon, or stripe + **gera** From Latin *generāre* –to generate.

Chapter 9

Meliponine Taxa:- Root Words, Prefixes & Suffixes

This section suggests root words, derivative words and suffixes used in the Meliponine Genus Taxon.

aca = Spanish acá - *(Used with a verb of motion)* **hither**, towards the speaker; **here**; Serbo-Croatian - Suffix **-ača** (*Cyrillic spelling* -ача) - Suffix appended to words to create a feminine noun, usually denoting a possession, performer, tool, place where the action occurs or a plant.

ae = English ae Variant of æ. - **æ** lower case (upper case Æ) - (chiefly archaic) A ligature of vowels a and e. Usage notes; Normally used for words of either Ancient Greek or Latin origin.; suffix -ae Representing both nominative and vocative plural case endings belonging to Latin first-declension feminine nouns. Usage notes: Although both -æ and -ae are normally used as plural forms in English, they are occasionally used as singular forms within the dative case or genitive case.

ai = Latin aī *or* āī - the second-person singular present active imperative of ***āiō****- present active* **āiō**, *present infinitive* āiere (deponent) - (defective) say, assert, say yes, say no, affirm, assent

ana = From Latin **ana** - old (of female persons and things), aged; Ancient Greek prefix '***ana-*** ' = up, on, again, apiece.; **From Latin suffix -ānus** m (*feminine* -āna, *neuter* -ānum); *first/second declension* - Of or pertaining to; -ian; usually indicates a relationship of position, possession, or origin.

api = *present active* **apō**, *present infinitive* apere, *perfect active* apī, *supine* aptum - fasten, attach, connect

apis = Latin **apis** f (*genitive* **apis**); *third declension* - A bee.;

arci = Latin – arcī - dative singular of arx = **arx** (*genitive* **arcis**); *f, third declension* - citadel, stronghold, height, hilltop;

aris = Latin *-āris*.(Suffix **-āris** m, f (*neuter* -āre); *third declension* - Used to form an adjective, usually from a noun, indicating a relationship or a pertaining to

aspis = From Ancient Greek *ἀσπίς* (aspis, “round shield or asp”).Latin **aspis** f (*genitive* aspidis); *third declension* - asp (*venomous snake*), viper

ata = From Latin Suffix ‘**-ata’** - Added to nouns to form adjectives Having, -ed.

atrata = From Latin **ātrātus** m (*feminine* ātrāta, *neuter* ātrātum); *first/second declension* - clothed in black (for mourning), darkened, blackened

atri = Latin atri - genitive masculine singular of āter (dark, black, gloomy)

atro = Latin **ātrō** dative masculine singular of *āter* - black, dark

auri = Latin **aurī** -genitive singular of *aurum* **-aurum** n (*genitive* aurī); *second declension* – gold-colour.

basis = Latin - An underlying condition or circumstance.

basi = Latin- dative singular of *basis* (*genitive* **basis**); f, *third declension* - A pedestal, foot, base; basis, foundation.

bi = From Latin *bis* (“twice”). Prefix ‘**bi-‘** = two

brevis = Latin - From Proto-Indo-European - m, f, breve n; *third declension* (of distance) small, little, narrow (of height or depth) short, low, shallow

caia = **Portuguese caia** - first-person singular present subjunctive of *cair* -From Old Portuguese *caer*, from Latin *cadere*, present active infinitive of *cadō*, from Proto-Indo-European **ḱad-* (“to fall”). Cognate with Spanish *caer*. **cair** (*first-person singular present indicative* caio, *past participle* caído) - to fall

cani = grey(Latin) **canī** - present passive infinitive of *canō* - **cānō** - dative masculine singular of *cānus* - **cānus** m (*feminine* cāna, *neuter* cānum); *first/second declension* - (of hair) gray

centri = **From Latin centrum** n (*genitive* centrī); *second declension* - centre/centre

cera = Related to Ancient Greek *κηρός* (kēros) - **Latin cēra** f (*genitive* cērae); *first declension* - wax, beeswax, honeycomb

cei = From Venetian **ceo** m (f **cea**, *m plural* **cei**, *f plural* **cee**) - small, minute

cephala = New Latin *cephala*, from Ancient Greek *κεφαλή* (kephalē, “head”)head**, cephalate** = zoology. Having a head.

ceps = New Latin “headed”

chaeta = English chaeta From **Ancient Greek *χαίτη*** (chaite, “crest, mane, flowing hair, foliage”) (*plural* **chaetae**) - A **chitinous bristle** of an **annelid worm** – English **chitinous** (*comparative* more **chitinous**, *superlative* most **chitinous**) - Made of, pertaining to, or resembling chitin.

chapad = Spanish **chapad** (*infinitive* chapar) - (Spain) Informal second-person plural (vosotros or vosotras) affirmative imperative form of chapar. - **chapar** (*first-person singular present* chapo, *first-person singular preterite* chapé, *past participle* chapado) to veneer, to plate

chelis = Latin chelis – meaning claws

cili = From Latin cili - chisel

clavi = Latin - dative singular of *clāvis*.- **clāvis** f (*genitive* clāvis); *third declension* - key, lever or bar for tightening a screw press

cocci = From Ancient Greek *κόκκος* (kokkos, “grain, seed, berry”). From Latin **coccum** n (*genitive* coccī); *second declension* - a scarlet berry of various plants The insect, *Coccus ilicis*, used for producing dye, a scarlet dye, or the cloth dyed with it

cola = Latin **cōlā - the** first-person singular present active imperative of *cōlō* - *present active* **colō**, *present infinitive* colere, *perfect active* coluī, *supine* cultum **-** till, cultivate., inhabit.; strain.

cornis = Latin adjective - m *and* f, **corne** n; *third declension* – horned

crassi = Latin **crassī -** nominative masculine plural of *crassus* - **crassus** m (*feminine* crassa, *neuter* crassum); *first/second declension* - Dense, thick, solid.

crus = Latin **crūs** n (*genitive* crūris); *third declension* - a leg

cup = From Old English *cuppe*, from Late Latin *cuppa*, probably a form of Latin *cūpa* (“tub”), from Proto-Indo-European **keup-* (“a hollow”) – English **cup** (*plural* cups) - A concave vessel for drinking from, usually made of opaque material (as opposed to a glass) and with a handle.

cutis = Latin cutis f (*genitive* **cutis**); *third declension* - living skin (insect cuticle)

dentata = From Latin **dentātus** m (*feminine* dentāta, *neuter* dentātum); *first/second declension* - toothed, having teeth

di = From Latin **deus** m (*genitive* deī); *second declension* (*nom. plural* deī or dī) - god, deity

dio = Latin diō - dative masculine singular of *dius* - **dius** m (*feminine* dia, *neuter* dium); *first/second declension* - Alternative form of dīvus. - **dīvus** m (*feminine* dīva, *neuter* dīvum); *first/second declension* - of or belonging to a deity; divine, godlike

do = Latin *present active* **dō**, - give. offer, render.

ea = Latin **eā** ablative feminine singular of *is*- **eā** (*not comparable*) - there, that way, on that side

Eo = alternative form eu- Ancient Greek *εὖ* (eu, "well, good") Prefix **eu- "**good, well, true, genuine."

eburnei = Borrowed from Latin *eburneus*. Italian **eburneo** m (*feminine* eburnea, *masculine plural* eburnei, *feminine plural* eburnee) - ivory (attributive)

emery = An impure type of corundum, often used for sanding or polishing

ella = Latin Suffix **'-ella'** nominative feminine singular of -ellus Suffix **-ellus** m (feminine -ella, neuter -ellum); first/second declension; Usage notes: The suffix -ellus is added to a noun to form a diminutive of that noun.

ensis = Latin Suffix **-ēnsis** m, f (*neuter* -ēnse); *third declension* - Of or from a place.

erythro- = From Ancient Greek *ἐρυθρός* ("red"). English Prefix **erythro-** used to form scientific terms meaning *red*, or showing a relationship to red blood

eu = English Prefix **eu-** = good, well, true, genuine

-eo = Latin suffix *present active* **eō**, *present infinitive* **īre**, *perfect active* **iī**, *supine* **itum** (irregular) - go, proceed, advance, set about.

fera = Latin - (*genitive* **ferae**); *f*, *first declension* -wild, beast

flavi = Latin (intransitive) to be yellow

flavo = Latin flāvō -dative masculine singular of *flāvus* - **flāvus** m (*feminine* flāva, *neuter* flāvum); *first/second declension* - yellow, golden, blond, flaxen

Frieseo = Named after the entomologist Friese + -eo (see suffix **–eo**)

frons = the upper anterior portion of an insect's head, above or behind the clypeus; forehead.

formis = Latin **fōrmīs** - dative plural of *fōrma* - **fōrma** f (*genitive* fōrmae); *first declension* - form, appearance

fulva = see **fulvo**

fulvi = Latin **fulvī** - nominative masculine plural of *fulvus*(see **fulvo**)

fulvo = Latin - fulvō dative masculine singular of fulvus - **fulvus** m (*feminine* fulva, *neuter* fulvum); *first/second declension* - reddish-yellow, tawny, amber-coloured, fulvous

fumi = From Latin fūmus m (*genitive* fūmī); *second declension* - smoke, steam

fusci = Latin fusci - nominative masculine plural of fuscus = m (*feminine* **fusca**, *neuter* **fuscum**); *first/second declension* - dark, black

fusco = make dark, swarthy or dusky; blacken, darken

gaster = From Ancient Greek *γαστήρ* (gastêr). Latin **gaster** f (*genitive* gasteris); *third declension* - belly

Genio– = prefix **English** Forms words relating to the chin

geneos = Latin geneos – class, race

hirsuti = From Proto-Indo-European **ghers-tu*, from **ghers* ("to bristle"), same root as *horreo* - **hirsūtus** m (*feminine* hirsūta, *neuter* hirsūtum); *first/second declension* - shaggy, bristly, prickly, **hairy,**

hirta = Latin – hirta nominative feminine singular of ***hirtus***- **hirtus** m (*feminine* hirta, *neuter* hirtum); *first/second declension* - hairy, shaggy

Addenda (New Names from 2015 not previously included)

Acknowledgements:

I am deeply grateful to Dr Claus Rasmussen for his advice and help on the etymology of most of the taxa herein. Without his advice and new key on the genera and subgenera of Indo-Malaya, Australasia and Papuasia, this study may not have been possible to complete.

Much appreciation is for Dr Anne Dollin for her advice and help in providing much information on Australian stingless bees in the Australasian taxa.

This author wishes to thank Dr Shashidhar Viraktamath of Department of Entomology, University of Agricultural Sciences, Bengaluru, Karnataka, INDIA for providing the etymology of his “Two New Species of Lisotrigona Moure” published in 2017 and “Two new species of Tetragonula (Hymenoptera: Apidae: Meliponini) from North-East India with notes on their nest structure” published by Institute of Zoology, Slovak Academy of Sciences in January 2020.

My gratitude also goes to Dr Michael S. Engel of the University of Kansas (2000–) Division of Invertebrate Zoology, American Museum of Natural History; Division of Entomology, Natural History Museum, and Department of Ecology and Evolutionary Biology, University of Kansas, Lawrence.& AMNH (1998–2000) for providing updated information on the latest taxa for Meliponini, Apidae, Hymenoptera and other corrections herein. Also included are new names from Stingless bees in Miocene amber of south-eastern China (Hymenoptera: Apidae) Journal of Melittology Bee Biology, Ecology, Evolution, & Systematics No. 105, pp. 1–83, 26 August 2021.

Names omitted in the first edition plus those published from 2015 to date.

Australia

magna, *Austroplebeia magna* Dollin, Dollin & Rasmussen 2015 [Etym.: The Latin feminine adjective, magna, meaning 'large', refers to the broad basitarsus III and long sting lancet in workers of this species.]

Remarks. The NT populations identified as *A. magna* sp. nov. were part of the 'symei' group defined in our preliminary studies (Halcroft et al., 2015). Its dark colouration characterised that group. It also included the eastern species, *A. cassiae.* Although superficially similar to *A. cassiae* in size and colouration, *A. magna* workers could be distinguished by their broad basitarsus III, long sting lancet and fine clypeus hair.

Reference:

Dollin, A. E.; Dollin, L. J.; Rasmussen, C. (2015-11-23). Australian and New Guinean Stingless Bees of the Genus Austroplebeia Moure (Hymenoptera: Apidae)—a revision. Zootaxa. 4047 (1): 1. doi:10.11646/zootaxa.4047.1.1. ISSN 1175-5334.

Indo-Malaya

Borneotrigona Engel, new subgenus, Type species: *Trigona hobbyi* Schwarz, 1937. **Etym**: The new subgeneric name is a combination of Borneo, the island from which the group is known, and Trigona Jurine. The gender of the name is feminine.

Included species: The subgenus presently includes only the type species, *H. (Borneotrigona) hobbyi* from northern Borneo [Brunei and Malaysia (Sarawak and Sabah)] (Engel 2019).

Reference:

Michael S. Engel1, Notes on Papuasian and Malesian stingless bees, with the descriptions of new taxa (Hymenoptera: Apidae) Journal of Melittology No. 88, pp. 1–25. 12 August 2019.

malaipanae, *Tetragonula (Tetragonula) malaipanae* Engel, Michener & Boontop 2017 [Etym: The specific epithet honours Savitree Malaipan of the Department of Agriculture, Kasetsart University, Bangkok, beloved mentor of Y.B. and a prominent Thai melittologist.] Holotype: Thailand, Kanchanaburi [Province], Thong Pha Phum [District], 17.ix.2007 [17 September 2007],

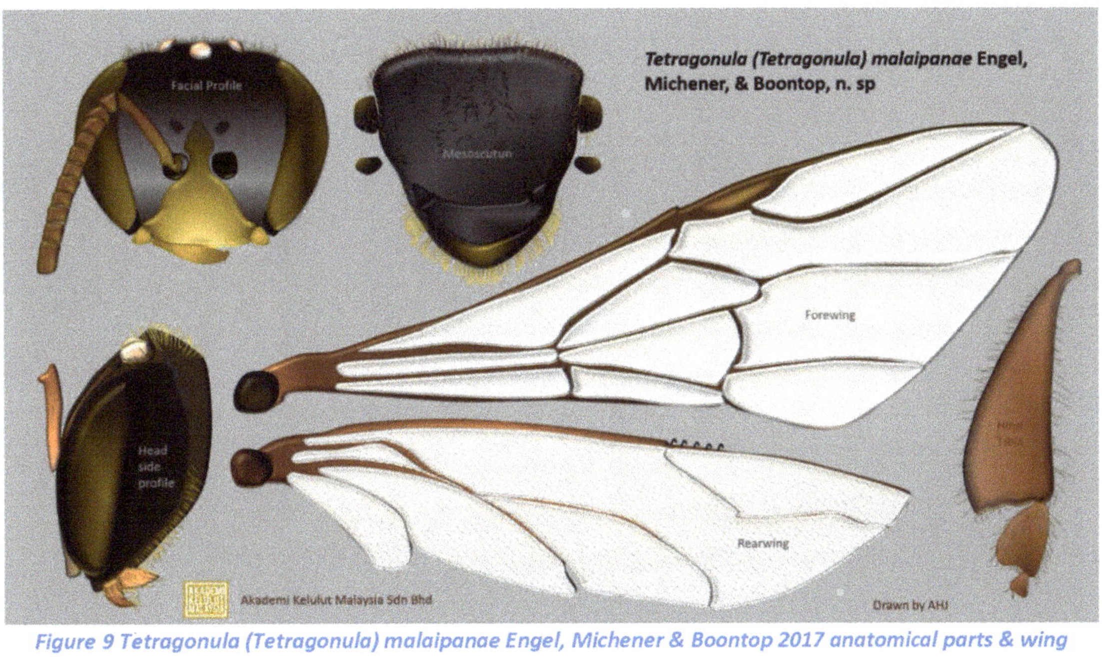

Figure 9 Tetragonula (Tetragonula) malaipanae Engel, Michener & Boontop 2017 anatomical parts & wing venation chart

satun, *Lepidotrigona* Attasopa and Bänziger n. sp. 2018 Etymology. The specific epithet refers to the province in Thailand where the species was collected;

Map 8 Province of Sarun near the Thailand – Malaysia border

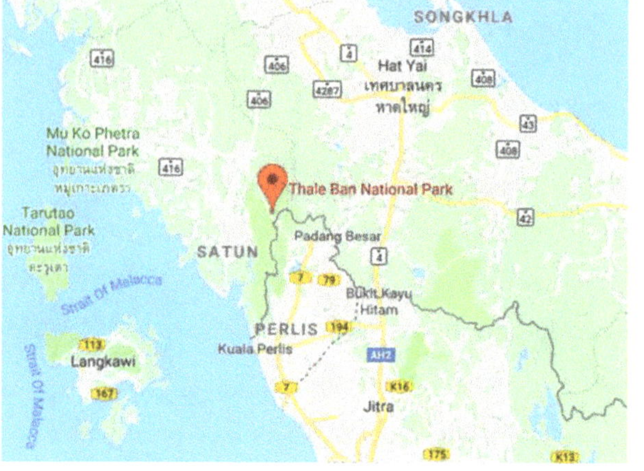

Map 9 Location of the *L. satun* collection site.

Figure 10 Comparison Lepidotrigona satun n. sp. worker to drone (Attasopa & Bänziger 2018)

Reference:

Engel M.S., Michener C.D., & Boontop Y., Notes on Southeast Asian Stingless Bees of the Genus Tetragonula (Hymenoptera: Apidae), with the Description of a New Species from Thailand. American Museum Novitates Number 3886, 17 pp. December 6, 2017.

Attasopa K., Bänziger H., Disayathanoowat T., & Packer L., A new species of Lepidotrigona (Hymenoptera: Apidae) from Thailand with the description of males of L. flavibasis and L. doipaensis and comments on asymmetrical genitalia in bees. Zootaxa 4442 (1): 063–082, 2018.

basimaculata, *Melipona* Bingham 1903* (Moure's Bee Catalogue, 2008) and is newly accepted in 2019 as *Tetragonula basimaculata* (Bingham, 1903) [TSN: 1129864 ITIS Bees Checklist] Etym: **basi** - Latin- **basis** - A pedestal, foot, base; + maculate (Latin –"spotted or stained"); having stained base or feet.

Notes: Biserat, Telom, Cameron Highlands. 15th **January, I *902.*** Resting on the ceiling at midday. 'This species resembles *M.* ***ventralis*** a little but is very much larger and with different sculptures. (Bingham 1903)

Figure 11 Redrawn from silhouette illustration in Bingham 1903.

Wallacea

Wallacetrigona Rasmussen et al. 2017 [Etym.: The generic name honours Alfred Russel Wallace (1823–1913), an intrepid and early explorer of the Indomalayan insect fauna and co-author with Charles R. Darwin (1809–1882) of the theory of evolution. The gender of the name is feminine.

Included species: Presently, the genus is understood to only include the type species, *Wallacetrigona incisa* (Sakagami and Inoue), a new combination from Indonesia (Sulawesi), east of the important faunal boundary known as the Wallace Line (and part of the biogeographic area known as Wallacea).

Reference:

Rasmussen C., Thomas J.C., & Engel M.S., A New Genus of Eastern Hemisphere Stingless Bees (Hymenoptera: Apidae), with a Key to the Supraspecific Groups of Indomalayan and Australasian Meliponini. American Museum Novitates Number 3888, 33 pp. December 14, 2017

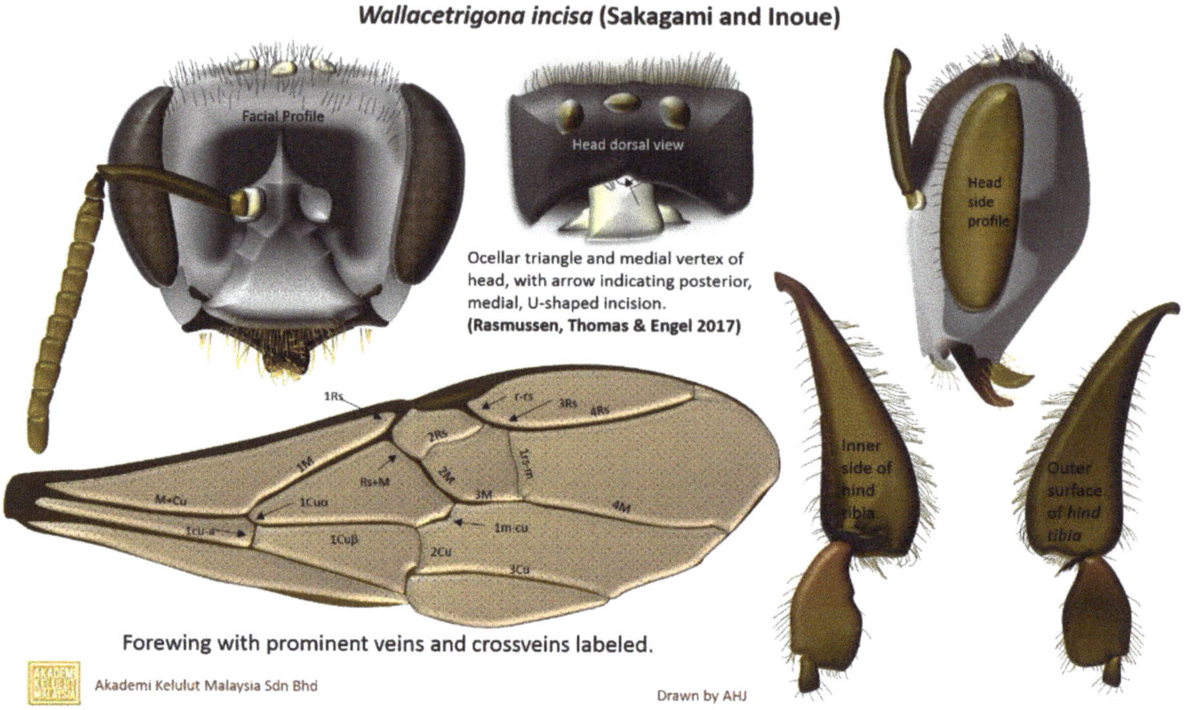

Figure 12 Wallacetrigona incisa (Sakagami & Inoue) anatomical parts and wing venation chart

Sahul

Sahulotrigona Engel & Rasmussen 2017 [Etym.: The subgeneric name is a combination of Sahul, a reference to the occurrence of the known species within a distinct zoogeographic region relative to those *Heterotrigona* occurring in Sundaland, and the generic name *Trigona* Jurine.]

Remarks: The gender of the name is feminine. Included species: The subgenus currently constituted contains only two species, both from New Guinea: *Heterotrigona (Sahulotrigona), paradisaea*, n. sp., and *H. (S.) atricornis* (Smith). (excerpts from A new subgenus of *Heterotrigona* from New Guinea (Hymenoptera: Apidae), Engel & Rasmussen 2017) Journal of Melittology No. 73 pg. 8.

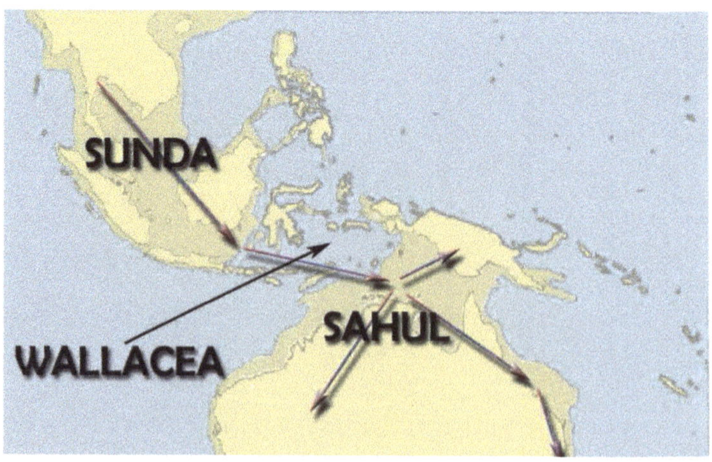

Map 10 Map of Sunda and Sahul - underwater continents

Map of Sunda and Sahul and the Wallace Line, the Lydekker Line and the Weber Line. Dated, 1 October 2012, 13:11:25 by Altaileopard.

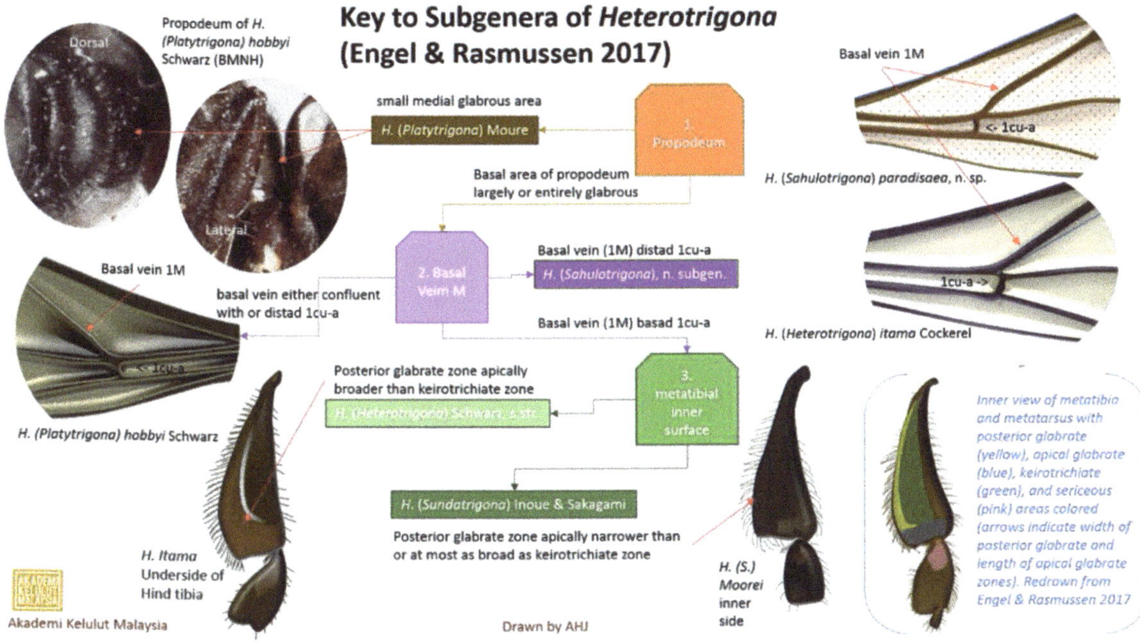

Figure 13 Composite Key of Heterotrigona (Engel & Rasmussen 2017)

paradisaea, *Heterotrigona (Sahulotrigona) paradisaea*

[Etym: The specific epithet is based on the generic name *Paradisaea* Linnaeus (Aves: Passeriformes: Paradisaeidae), the genus of the Emperor bird-of-paradise (*Paradisaea guilielmi* Cabanis), which is the national emblem of Papua New Guinea.]

Figure 14 Emblem of Papua New Guinea

Note: The name is meant to honour the unique biota cultivated by the indigenous peoples of New Guinea.

Reference:

Michael S. Engel & Claus Rasmussen, A new subgenus of Heterotrigona from New Guinea (Hymenoptera: Apidae) Journal of Melittology No. 73, pp. 1–16 3 November 2017

Figure 15 Heterotrigona (Sahulotrigona) paradisaea Engel & Rasmussen 2017 anatomical parts and wing venation

tricholoma, *Heterotrigona (Sahulotrigona)* Engel, new species [Etym.: The specific epithet is taken from the Greek words, trichós (τρῖχός), meaning "hair", and lôma (λῶμᾰ), meaning, "border" or "fringe", and is a reference to the fringe of setae bordering the mesoscutum posterolaterally.]

taraxis, *Heterotrigona (Sahulotrigona)* Engel, new species [Etym.: The specific epithet is taken from the Greek, *taraxis* (τᾰρᾰχῆς), meaning, "confusion", and is a reference to the historical confusion regarding the identity of this species.]

Comments: This species has been long confused as "*Trigona atricornis* Smith" a recent examination of the holotype of *T. atricornis* in the Oxford University Museum of Natural History reveals that Smith's (1865) species in actuality belongs to the genus *Papuatrigona* Michener & Sakagami, wherein it is a senior synonym of *T. genalis* Friese (*vide infra*) (Engel 2019).

Reference:

Michael S. Engel1, Notes on Papuasian and Malesian stingless bees, with the descriptions of new taxa (Hymenoptera: Apidae) Journal of Melittology No. 88, pp. 1–25. 12 August 2019.

Indian Subcontinent

Flavotetragonula, Etym: The subgeneric name is derived from flavus in Latin for yellow. [(1)]

Abstract: A new subgenus of stingless bees, *Flavotetragonula* Shanas, subgen. n. is established, and three new species, *Tetragonula (Flavotetragonula) calophyllae* Shanas and Faseeh, n. sp., *Tetragonula (Tetragonula) perlucipinnae* Faseeh and Shanas, n. sp. and *Tetragonula (Tetragonula) travancorica* Shanas and Faseeh, n. sp. is described from southern India, based on workers.

Association for Advancement of Entomology 2019

DOI: 10.33307/entomon.v44i1.424

calophyllae - *Tetragonula (Flavotetragonula) calophyllae* Shanas and Faseeh, n. sp. [(1)], **Etym**: The specific epithet is after the generic name of the endangered tree Calophyllum inophyllum, on which the first feral colony was observed.

perlucipinnae - *Tetragonula (Tetragonula) perlucipinnae* Faseeh and Shanas, n. sp. [(1)] **Etym**: The specific epithet is based on the Latin words perlucidulus and pinna, alludes to the transparent wing.

travancorica -*Tetragonula (Tetragonula) travancorica* Shanas and Faseeh, n. sp.[(1)], **Etym**: This species is named after the erstwhile kingdom of Travancore.

Remarks: The Kingdom of Travancore (/ˈtrævənkɔːr/) (Thiruvithamkoor) was an Indian kingdom from c. 870 until 1949. It was ruled by the Travancore Royal Family from Padmanabhapuram and later Thiruvananthapuram. At its zenith, the kingdom covered most modern-day central and southern Kerala with the Thachudaya Kaimal's enclave of Irinjalakuda Koodalmanikyam temple neighbouring Kingdom of Cochin, as well as the district of Kanyakumari, now in the Indian state of Tamil Nadu.

chandrai, *Lisotrigona chandrai* Viraktamath and Sajan Jose 2017[(2)] [syn. to *Lisotrigona cacciae* (Nurse 1907)] (Rasmussen 2017) [Etym: Dedicated to Indian taxonomist Dr Chandrashekhar A Viraktamath]

Remarks: This is named after a well-known Indian Taxonomist Dr Chandrashekhar A Viraktamath, popularly known as "Chandra" among his international colleagues and friends. Dr CAV (that is how he is known among his Indian students and colleagues) has completely dedicated his life to the study of Indian insects, especially leafhoppers. He is a mentor and guides a host of Indian and International students, including the authors of this paper. Shortly he is turning 75 and suffering from failing eyesight, but still, he works for 18 hours a day in his laboratory and home. Hence, this new species of stingless bees is named as a tribute to this great Indian personality. (see **Note)

revanai, *Lisotrigona revanai* Viraktamath and Sajan Jose 2017(2) [Etym: The species is named after a famous Hindu God, "Revanasiddeshwara". syn. to *Lisotrigona cacciae* (Nurse 1907)] (Rasmussen 2017). Type Loc. It is collected in Sawantawadi, Maharashtra, India. (Pers. Comm. Viraktamah, 2017)]

Remarks: The species is named after a famous Hindu God, "Revanasiddeshwara". The temple is located near the type locality of the new species on a hillock known as "Revanadi" in Maharashtra. Also known as a **Theonym** - A **theonym** (from classical Greek *theos* "god" and *-onym* "name") is a proper name of a deity.

kyrdemkulaiensis, *Tetragonula kyrdemkulaiensis* Viraktamath and Rojeet sp. n.(3) Etym: This species is named after the place where it is collected. Holotype: Male: Meghalaya: Kyrdemkulai (25.44° N, 91.47° E, Altitude 679 m a.s.l.), 20. ii. 2019, Coll. Rojeet T., deposited at UASB.

Remarks: This stingless bee was found in a subterranean nest at Kyrdemkulai, Ri-Bhoi District of Meghalaya, in January 2019. The nest was located at a depth of 90 cm in the soil with an entrance burrow (116 cm in length) running horizontally.

srikantanathi, *Tetragonula srikantanathi* Viraktamath sp. n. (3) This species is named in honor of Mr. Srikanta Nath. Holotype: Female: Tripura: Salema (24.01° N 91.83° E, Altitude 89 m a.s.l.), 8. xi. 2017, Coll. S. Viraktamath, deposited at UASB.

Remarks: Mr. Srikanta Nath, Assistant Director of Agriculture of the Tripura State Agriculture Department who searched the nest and helped in studying its structure.

Reference for New Names up to January 2020:

(1)Shanas S. & Faseeh P., A new subgenus and three new species of stingless bees Hymenoptera: Apidae: Apinae: Meliponini) from India. Entonom Vol 44 No 1 (2019)

(2)Viraktamath S. & Jose K. S., Two New Species Of Lisotrigona Moure (Hymenoptera: Apidae: Meliponini) From India With Notes On Nest Structure. The Bioscan 12(1): 21-28, 2017

(3)Viraktamath S. & Rojeet T., Two new species of Tetragonula (Hymenoptera: Apidae: Meliponini) from North-East India with notes on their nest structure. Institute of Zoology, Slovak Academy of Sciences, 20 January 2020

Argentina

A new species of *Plebeia* Schwarz (Hymenoptera: Apidae) from Argentina is described and illustrated. *Plebeia mansita* Alvarez & Rasmussen sp. nov. occurs in the extreme northwest of the country and is restricted to the biogeographic province of the Yungas, where it is found at altitudes above 1,000 m.

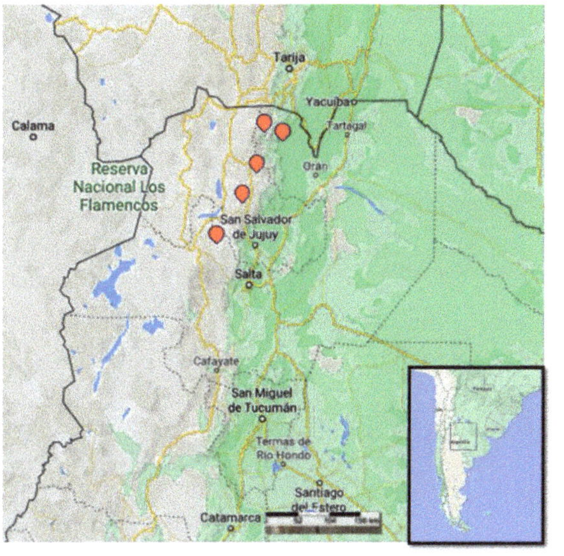

Map 11 Distribution map of *Plebeia mansita* sp. nov.

mansita, *Plebeia* Alvarez & Rasmussen sp. nov. [Etym.: The epithet refers to the common name used by local people in Baritú, El Lipeo and Los Toldos, "mansita", due to the docile behaviour of the bees when disturbed or handled.]

Remarks: Species of *Plebeia* are generally very abundant and occur in most of the Neotropical region, from Nuevo León, Mexico, to Buenos Aires and San Luis, Argentina, also representing the distributional boundary of the stingless bee tribe in the Neotropical region (Roig-Alsina & Alvarez, 2017).

Reference: www.biotaxa.org/RSEA. ISSN 1851-7471 (online)

Alvarez L. J., Flores F. F. & Rasmussen C., A new species of Plebeia Schwarz (Hymenoptera: Apidae) from the Argentine Yungas Revista de la Sociedad Entomológica Argentina 80(2): 35-41, 2021

Brazil

Map 12 Serra da Bocaina, in São Paulo & Chapada dos Veadeiros, in Goiás.

Two new species of the stingless bee genus *Schwarziana* from Brazil are described and illustrated. *Schwarziana bocainensis* sp. nov. is described from Serra da Bocaina, in São Paulo, and *S. chapadensis* sp. nov. is described from Chapada dos Veadeiros, in Goiás. An identification key to workers of the known species of Schwarziana is provided.

bocainensis - *Schwarziana bocainensis* Melo 2015 sp. nov. [Etym.: The species is named in reference to its type locality, the Serra da Bocaina in São Paulo, Brazil]

chapadensis - *Schwarziana chapadensis* sp. nov. is described from Chapada dos Veadeiros, in Goiás.

Reference: http://dx.doi.org/10.1016/j.rbe.2015.08.001

Gabriel A.R.Melo, New species of the stingless bee genus *Schwarziana* (Hymenoptera, Apidae) Rev. Bras. Entomol. vol.59 no.4 São Paulo Oct./Dec. 2015 Zootaxa 2015 May 11; 3956(3):389-402. doi:10.11646/zootaxa.3956.3.4.

camargoiana, *Trichotrigona camargoiana* Pedro and Cordeiro, 2015 sp. nov. Etym.: The species here described is named honouring the late Prof. Dr. João Maria Franco de Camargo (1941-2009), *Trichotrigona camargoiana* sp. nov. from Candeias do Jamari, Rondônia, Brazil.

Remarks: The second species of the rare and unusual genus *Trichotrigona* Camargo & Moure, 1983. The new species differs from *T. extranea* Camargo & Moure, 1983 by having plumose setae on the face and mesepisterna, wing setae predominantly yellow and black metatibia. The diagnosis for *Trichotrigona* is updated. As believed for *T. extranea, T. camargoiana* sp. nov. does not store food and is a cleptobiotic social bee associated with Frieseomelitta.

Reference:

Pedro, Silvia R. M., Cordeiro, Guaraci Duran (2015): A new species of the stingless bee Trichotrigona (Hymenoptera: Apidae, Meliponini). Plazi.org taxonomic treatments database. Checklist dataset https://doi.org/10.11646/zootaxa.3956.3.4 accessed via GBIF.org on 2021-04-20

Africa

transvaalensis, *Trigona transvaalensis* Cameron, 1914 NHMUK: ecatalogue:9090631

Remarks: Transvaal is a historical geographic term associated with land north of (i.e., beyond) the Vaal River in South Africa. Several states and administrative divisions have carried the name Transvaal. The Transvaal's name, which means "across the Vaal," originated with the Afrikaners who, in the 1830s, migrated to the region after crossing the Vaal River.

mimica. *Melipona alinderi* var. *mimica* Alfken, 1932. Kenya - Latin mimica - accusative neuter plural of mīmicus From Ancient Greek μῑμικός (mīmikós). mīmicus (feminine mīmica, neuter mīmicum); first/second-declension adjective 1. mimic, 2. farcical

somalina, Trigona *beccarii* var. *somalina* Magretti 1898. [Etym.: Named after Somali Region of Ethiopia, where Sankurar is located] Magretti 1898b: 586, worker holotype (MCSNG). Type locality: Ethiopia, "Sancurar e gli Amarr", II-IV.1896, leg. Bottego (Sankurar, 4°01'N 40°07'E; Amarr, ca 5°10'N 36°40'E). = Somali + suffix '-ina'

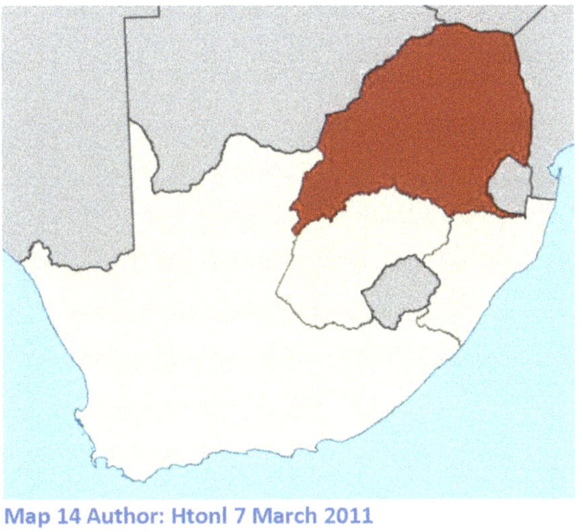

Map 14 Author: Htonl 7 March 2011 https://commons.wikimedia.org/wiki/File:Map_of_the_provinces_of_South_Africa_1976-1994_with_the_Transvaal_highlighted.svg

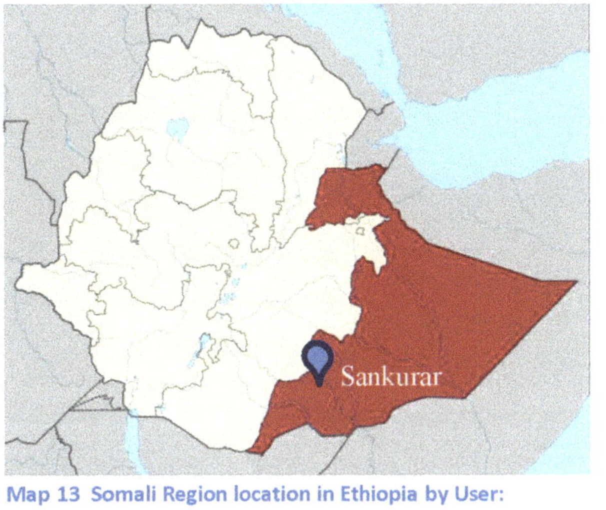

Map 13 Somali Region location in Ethiopia by User: AlaskaLava 9 July 2017 https://commons.wikimedia.org/wiki/File:Somali_Region_Map.svg

Remarks: The regional state borders the Ethiopian states of Afar and Oromia and the chartered city Dire Dawa (Dire Dhawa) to the west, as well as Djibouti and Somaliland to the north and northeast, Somalia to the east and south; and Kenya to the southwest.

Reference:

Eardley C. D., Taxonomic revision of the African stingless bees (Apoidea: Apidae: Apinae: Meliponini) African Plant Protection vol. 10, no. 2, August 2004. http://zoologie.umons.ac.be/hymenoptera/biblio/01000/eardley2004_3234.pdf

voeltzkowi, *Liotrigona* (de Saussure, 1890) Madagascar. Pauly et al. 2001: 334. [Etym: The specific epithet honours **Alfred Voeltzkow** (14 April 1860 – February 1947)]

Alfred Voeltzkow was a German zoologist and botanist from Berlin, Brandenburg. He is remembered for his scientific journeys in eastern Africa and the Western Indian Ocean; (Comoros, Madagascar, Juan de Nova and Aldabra).

https://en.wikipedia.org/wiki/Alfred_Voeltzkow

New Names as up to August 2021

florilega, *Tetragonula* (*Tetragonula*) *florilega* Engel, new species; **Type locality**: Zhangpu County, Zhangzhou Prefecture, Fujian Province, China.; Horizon: Middle Miocene: Langhian, 14.7 Ma; sedimentary layer II, Fotan Group.; **Etymology**: The specific epithet is taken from the Latin florilegus, meaning "flower-gathering."

Anteplebeina Engel, new subgenus; **Type species**: *Austroplebeia* (*Anteplebeina*) *fujianica* Engel, new species.; **Etymology**: The new subgeneric name is a combination of the Latin words ante ("before"), plēbēia ("plebeian", "commoner", "plebe"), and the feminine diminutive, suffix –īna ("wife of," feminine suffix with diminutive implication). The gender of the name is considered feminine.

fujianica, *Austroplebeia* (*Anteplebeina*) *fujianica* Engel, new species.; **Type locality**: Zhangpu County, Zhangzhou Prefecture, Fujian Province, China.; **Horizon**: Middle Miocene: Langhian, 14.7 Ma; sedimentary layer II, Fotan Group.; **Etymology**: The specific epithet is taken from Fujian Province in southeastern China, the region encompassing the Zhangpu locality.

Atrichotrigona Engel, new subgenus.; **Type species**: *Trigona cameroonensis* Friese, 1900.; **Etymology**: The new subgeneric name is a combination of the Greek alpha privativum, a– (ἀ–, indicating negation or opposite), trikhós (τρῐχός, meaning, "hairy"), and the generic name Trigona [tri– (τρῐ–, meaning, "three") and gōnía (γωνία, meaning, "corner" or "angle")]. The gender of the name is feminine.; Included species: Aside from the type species, the subgenus includes *Axestotrigona simpsoni* Moure.

Lispotrigona V.H. Gonzalez & Engel, new subgenus.; **Type species**: *Trigona schultzei* Friese, 1901.; **Etymology**: The new subgeneric name is a combination of the Greek *líspos* (*λῐ́σπος*, meaning, "polished") and the generic name *Trigona*. The gender of the name is feminine.; **Included species**: Aside from the type species, the subgenus includes *Nannotrigona* (*Lispotrigona*) *dutrae* (Friese).

Asperplebeia Engel, new genus.; **Type species**: *Plebeia moureana* Ayala, 1999.; **Etymology**: The new genus-group name is a combination of the Latin terms *asper* (meaning, "rough" or "coarse") and *plēbēia* (meaning, "commoner"), in reference to the roughened and matte mesoscutum relative to *Plebeia*. The gender of the name is considered to be feminine.; **Included species**: Aside from the type species, *Asperplebeia moureana* (Ayala), the genus includes *A. tica* (Wille), both new combinations.

Genus *Plebeia* Schwarz

Nanoplebeia Engel, new subgenus.; **Type species**: *Trigona minima* Gibodo, 1893.; **Etymology**: The new genus-group name is a combination of the Latin terms *nānus* (meaning "dwarf") and *plēbēia* (meaning "commoner"). The gender of the name is considered to be feminine.; **Included species**: The subgenus currently includes *Plebeia franki* (Friese), *P. margaritae* Moure, *P. minima* (Gribodo), and at least four undescribed species (Engel, in press-d).

Genus *Schwarziana* Moure

Chapadapis Engel, new subgenus.; **Type species**: *Schwarziana chapadensis* Melo, 2015.; **Etymology**: The new subgeneric name is a combination of chapada, referring to the plateaus of the Brazilian Highlands, in reference to the Chapada dos Veadeiros, and *apis* (Latin, meaning, "bee"). The gender of the name is feminine.

Trigona Genus Group

Genus *Trigona* Jurine

Aphaneuropsis Engel, new subgenus.; **Type species**: *Centris cilipes* Fabricius, 1804.; **Etymology**: The new subgeneric name is a combination of Aphaneura Gray [aphanḗs (ἀφᾰνής, meaning, "unseen") and neûron (νεῦρον, meaning, "nerve")] and the Greek ópsis (ὄψῐς, meaning, "appearance"). The gender of the name is feminine.; **Included species**: Aside from the type species, the subgenus includes *T. lacteipennis* Friese, *T. mazucatoi Almeida,* and *T. pellucida* Cockerell.

Koilotrigona Engel, new subgenus.; **Type species**: Trigona fulviventris Guérin-Méneville, 1844.; **Etymology**: The new subgeneric name is a combination of the Greek koílos (κοῖλος, meaning, "hollow" or "concave") and Trigona. The gender of the name is feminine.; **Included species**: In addition to the type species, the subgenus currently includes *T. braueri* Friese and *T. guianae* Cockerell.

Necrotrigona Engel, new subgenus.; **Type species**: *Trigona hypogea* Silvestri, 1902.; **Etymology**: The new subgeneric name is a combination of the Greek nekrós (νεκρός, meaning, "dead") and Trigona. The gender of the name is feminine.; **Included species**: Aside from the type species, the subgenus includes *T. crassipes* (Fabricius) and *T. necrophaga* Camargo & Roubik.

Dichrotrigona Engel, new subgenus.; **Type species**: *Trigona dimidiata* Smith, 1854.; **Etymology**: The new subgeneric name is a combination of the Greek díkhrōs (δίχρως, meaning, "bicoloured") and Trigona. The gender of the name is feminine.; **Included species**: Aside from the type species, the subgenus currently includes *T. sesquipedalis* Almeida and *T. venezuelana* Schwarz.

Nostotrigona Engel, new subgenus.; **Type species**: *Trigona recursa* Smith, 1863.; **Etymology**: The new subgeneric name is a combination of the Greek nóstos (νόστος, meaning, "returning") and Trigona. The gender of the name is feminine.; **Included species**: In addition to the type species, the subgenus includes *T. permodica* Almeida.

Ktinotrofia Engel, new subgenus.; **Type species**: *Trigona fuscipennis* Friese, 1900.; **Etymology**: The new subgeneric name is a combination of the Greek ktinotrofia (κτηνοτροφία), meaning "animal husbandry", and refers to the obligate hemipterophily of these species. The gender of the name is feminine.; **Included species**: In addition to the type species, the subgenus includes *T. albipennis* Almeida.

Reference for New Names up to August 2021:

Michael S. Engel, et al. Stingless bees in Miocene amber of south-eastern China (Hymenoptera: Apidae) Journal of Melittology Bee Biology, Ecology, Evolution, & Systematics No. 105, pp. 1–83, 26 August 2021.

★Glossary ★

Terms Pertaining to Zoological Nomenclature

Allotype

A specimen designated from the type series is the opposite sex of the holotype.

Available name

A name that is correctly proposed according to the International Code of Zoological Nomenclature. An available name is not necessarily a valid name.

Category

Any rank within the classification hierarchy, e.g., family, subfamily, subspecies.

Change of rank

When a name is moved from one level of the classification system to another, e.g., when De Lotto (1955) moved ***Ceroplastes destructor brevicauda*** from the subspecies to the species rank ***C. brevicauda,*** this was a change of rank.

Classification

A system of nested hierarchical categories is used to efficiently store information about the diversity of life.

Classify

To place a taxon in a classification system.

Cotypes

A term no longer recognized in the International Code of Zoological Nomenclature; synonymous with syntypes.

Emendation

An intentional change to a previously proposed name, e.g., Lindinger proposed the emendation ***Hemiberlesea*** for the armoured scale ***Hemiberlesia,*** indicating that it was originally improperly formed.

Holotype

A specimen that serves as the standard-bearer of a species or subspecies name.

Homonym

One of two or more scientific names that are identical but pertain to different organisms, e.g., ***Eriococcus mancus*** Ferris, 1955 and ***Eriococcus mancus*** (Maskell, 1897); ***Onceropyga*** Ferris, 1955 and ***Onceropyga*** Turner, 1904.

Incertae sedis

A name of uncertain identity.

Junior homonym

If there are only two homonyms, the junior homonym is the most recently described homonym; if there are more than two homonyms, the junior homonyms are all but the oldest described homonym, which is the senior homonym, e.g., ***Eriococcus mancus*** Ferris, 1955 is the junior homonym, and ***Eriococcus mancus*** (Maskell, 1897) is the senior homonym.

Junior synonym

If there are only two synonyms, the most recently described one is the junior synonym; if there are more than two synonyms, the junior synonyms are all, but the oldest described one, which is the senior synonym, e.g., ***Apiomorpha nux*** Fuller, 1896 is the junior synonym and ***A. pharetrata*** Scharder, 1863 is the senior synonym.

Justified emendation

An emendation that is correct according to the International Code of Zoological Nomenclature, e.g., the name ***susani*** is proposed as a patronym for a woman named Susan; according to the Code, the name must be changed to ***susanae*** and is a justified emendation.

Lectotype

A specimen is chosen as the standard-bearer of a species or subspecies and selected from the syntype series.

Misidentification

A citation of a name in the literature that used the incorrect name because the specimens were improperly determined.

Misspelling

A citation of a name in the literature that is incorrectly spelt.

Neotype

A specimen was chosen as the standard-bearer of a species or subspecies name for which none of the original type specimens exists.

New combination

When a species is transferred to a different genus for the first time.

Nomen nudum

A name that does not fulfil the criteria set by the International Code of Zoological Nomenclature as a legally described scientific name and therefore cannot be used unless it is subsequently proposed correctly.

Nomen obltum

Latin for "forgotten name" This is a name that has not been used in the scientific community since 1899, and when it is either a senior synonym or a homonym. Once it has formally been declared to be a *nomen oblitum*, the disused name is to be 'forgotten'. By the same act, the junior name must be declared to be a *nomen protectum*; from then on, it takes precedence. Example: ***Aspidiotus rosae*** Bouché, 1833, is a *nomen protectum* taking precedence over its senior synonym ***Coccus rosae*** Virey, 1821 (*nomen oblitum*). See Article 23.9.2 of ICZN. See *nomen protectum* below.

Nomen protectum

Latin for “protected name" This is a name that has been given precedence over its unused senior synonym or senior homonym. See *nomen oblitum* above.

Paralectotype

All of the specimens in the syntype series of a species or subspecies other than the lectotype.

Paratype

All of the specimens in the type series of a species or subspecies other than the holotype.

Replacement name

A name that is assigned to replace a name that is a junior homonym, e.g., ***Onceropyga*** Turner, 1904 is the valid name, and ***Onceropyga*** Ferris, 1955 is the junior homonym and must be replaced; Hoy (1963) proposed the replacement name ***Oregmopyga***.

Senior homonym

The oldest described homonym, e.g., ***Onceropyga*** Turner, 1904 is the senior homonym and ***Onceropyga*** Ferris, 1955 is the junior homonym.

Senior synonym

The oldest synonym, e.g., ***Apiomorpha pharetrata*** Scharder, 1863 is the senior synonym and ***A. nux*** Fuller, 1896 is the junior synonym.

Synonym

One of two or more scientific names that are spelt differently but refer to the same organism, e.g., ***Apiomorpha nux*** Fuller, 1896 and ***A. pharetrata*** Scharder, 1863 are names used for the same species of eriococcid and are synonyms.

Synonymy

A section of a systematic presentation about an organism that lists all of the names that have been used for the organism, including synonyms, new combinations, misidentifications, etc. In some cases, this section may include only true synonyms.

Syntypes

The series of specimens used to describe a species or subspecies when the author did not include a holotype.

Systematics

The field of science dealing with the diversity of life and the relationships of life's component organisms.

Taxon

One or more organisms belong to the same taxonomic unit.

Taxonomy

The field of science classifies life.

Topotype

One or more specimens are collected at the same location as the type series regardless of whether they are part of the type series.

Type

A term used to describe the nomenclatural importance of various kinds of specimens.

Type locality

The geographic location where the primary type was collected.

Type species

A species that has been selected as the standard-bearer of a genus or subgenus.

Type genus

A genus that has been selected as the standard-bearer of a tribe, family, or superfamily and provides the stem of the family-group name.

Unavailable name

A name that is incorrectly proposed according to the International Code of Zoological Nomenclature.

Unjustified emendation

An emendation that is incorrect according to the International Code of Zoological Nomenclature, e.g., the generic name ***Hemiberlesea*** Lindinger is an incorrect change of ***Hemiberlesia*** Cockerell according to the Code and is an unjustified emendation.

Valid name

The correct name of an organism, e.g., if ***Apiomorpha nux*** Fuller, 1896 and ***A. pharetrata*** Scharder, 1863 apply to the same species (and therefore are synonyms), then by the law of priority (the oldest name prevails) ***A. pharetrata*** Scharder, 1863 is the valid name.

Taxonomy glossary

epithet - An **epithet** (from Greek: ἐπίθετον *epitheton*, neut. of ἐπίθετος *epithetos*, "attributed, added"[1]) or **byname** is a descriptive term (word or phrase) accompanying or occurring in place of a name and having entered common usage. It can be described as a glorified nickname. It has various shades of meaning when applied to seemingly real or fictitious people, divinities, objects, and binomial.

homonym = English (*homo-* + *-onym)* - **homonym** (*plural* **homonyms**)

1. (loosely) A word that sounds or is spelt the same as another word but has a different meaning is technically called a *homophone* (same sound) or a *homograph* (same spelling).
2. (taxonomy) A name for a taxon that is identical in spelling to another name that belongs to a different taxon.

incertae setae = (Latin for "of uncertain placement") is a term used to define a taxonomic group where its broader relationships are unknown or undefined. Alternatively, such groups are frequently referred to as "enigmatic taxa". Uncertainty at specific taxonomic levels is attributed by *incertae familiae* (of the uncertain family), *incerti subordinis* (of uncertain suborder), *incerti ordinis* (of uncertain order) and similar terms.

lapsus = Latin **lāpsus** m (*genitive* **lāpsus**); *fourth declension* - gliding, sliding, slipping, falling

nomen novum = In biological nomenclature, a ***nomen novum*** (Latin for "new name"), **new replacement name** (or replacement name, new substitute name, substitute name[1]) is a technical term. It indicates a scientific name that is created specifically to replace another scientific name, but only when this other name can not be used for technical, nomenclatural reasons (for example, because it is a homonym: it is spelt the same as an existing, older name); it does not apply when a name is changed for taxonomic reasons (representing a change in scientific insight).

nomen ad interim = meaning a provisional name. The Latin phrase ***ad interim*** (abbr. *ad int*, literally "in the time between") means "in the meantime" or "temporarily".

sensu = Latin **sensū** - ablative singular of *sēnsus*-**sēnsus** m (*genitive* **sēnsūs**); *fourth declension*

1. perception, the capability of feeling, ability to perceive
2. a feeling, sentiment
3. (poetic) understanding, reason

lato = Latin dative masculine singular of *lātus*- lātus m (*feminine* **lāta**, *neuter* **lātum**); *first/second declension*

1. born, carried, having been carried
2. suffered, endured, having been suffered
3. reported, having been reported

sensu lato = From the ablative case of *sēnsus* ("sense") and *lātus* ("broad") - **sēnsū lātō** (*not comparable*)- In the broad sense.

sensu stricto = From Latin *sēnsū strictō* ("in the narrow sense"). (taxonomy) (Often abbreviated: ***s.s.***, ***s. str.***, ***sens. str.*** *or* ***sens. strict.****) In the narrow sense. Added after a taxon to mean the taxon is being used in the sense of the original author, or without taxa which may otherwise be associated with it.*

sic = Latin **sīc** (*not comparable*)

1. thus, so, or just like that
2. yet

supra = Original form *superā*. Related to Ancient Greek *ὑπέρ* (hyper, "above"), a prefix to indicate higher grade/rank, literally meaning over. Cf. *super*. - **suprā** (*not comparable*)

1. (of place) above, on the top, on the upper side
2. (of time) before, previously, formerly

Derived terms

- supra quam, (rarer) supra quod (above or beyond what, more than)

Usage notes

- When pertaining to time, it especially refers to anything previously said or written. Also, please note that Lewis & Short states that the first syllable is long as *sūprā*.

Preposition

suprā (+ *accusative*)

1. (of location) over, above, beyond

__Supra__ naturam.

Above nature.

2. (of time) before

   __Supra__ septingentesimum annum.

   Paulo __supra__ hanc memoriam.

3. (of number, degree, or quantity) over, above, beyond, more than [syn.]
4. (of employment or office) over, in authority over, in charge of

   Quos __supra__ somnum habebat.

Derived terms

- suprā caput sum (I am close at hand)

Composition Of Scientific Words

Abbreviations And Symbols

Ab.Am Aboriginal American
Abyss Abyssinian
Ar Arabic
AS Anglo-Saxon
C Celtic
Chin Chinese
D Dutch
Dan Danish
E English
Egypt Egyptian
F French
G German
Goth Gothic
Gr Greek
Heb Hebrew
Hind Hindi
Ir Irish
It Italian
Jap Japanese
L Latin
LL Late Latin
ME Middle English
MHG Middle High German
ML Medieval Latin
X Norse
XL New Latin

OE Old English
OF Old French
OHG Old High German
OX Old Xorse
Per Persian
Pg Portuguese
Puss Russian
Skt Sanskrit
Scan Scandinavian
Slav Slavic
Sp Spanish
Sw Swedish
Teut Teutonic
Turk Turkish
VL Vulgar Latin
W Welsh
ant antonym
c common
cf compare
chem chemistry
cog cognate
comp comparative
cont continued
der derived
dim diminutive
e.g. for example
etc. et cetera
etym etymology

ex example
exc exception
f feminine
gen genitive
lit literally
m masculine
n neuter
nom nominative
part participle
perf perfect
pi plural
pref prefix
prob probably
q.v which see
sing singular
suf suffix

super superlative
syn synonym
< derived from
? doubtful, probably
[ ] brackets enclose editorial assumptions, corrections, and opinions

= equals

Table 3 Old Tupi Language Glossary

Reported here, different names of stingless bees inside the interior Northeast.

No confirmation, but as a curiosity, it is worth knowing: popular accounts, without knowledge.

Note: these species must have already been identified by description are known to scholars.

Cuviara - indigenous bees that lodge in the sticks feet are medium size and aggressive type curls in hair and skin ordisca (perhaps bee dog)

Feri Feri - indigenous bee, round head, black thorax and abdomen black coffee, black striped, transparent wings - other names for this bee - Brabo - straw

Frecheira - Bee indigenous small, measuring 4mm, head and thorax black, matte orange-red abdóeme short, clear wings, nests in bounds and stone walls, honey tart.

Guarili - Bee indigenous large, measuring 9mm, all orange, striped black, Arua in Guaraní, could not confirm this species.

Jandira - Bee indigenous Northeast, measures 11mm, all black, white striped, entry has a wide mouth, wild honey flavour, like flower and eucalyptus.

Mangava white - bee Northeast, very light yellow, with black stripes, say it is kind of marauding black, does not work, sometimes live in the same hole, the esp [ECIES black.

Montari - Bee indigenous Northeast, measures 13mm, matte black head and thorax, abdomen striped branco.Uma popular name referring to annatto.

Tuvininha - Bee indigenous unknown, but the name is suggestive rhymes with Tujuvinha, group of junior, so a nickname from the same regional.

Tuvuna - Bee indigenous unknown, perhaps a nickname regional tubuna

Curupireira - the name of indigenous honey bee-related toxic Cupira are bees
black, aggressive smell ocher (coconut tart)

Tuiva - information Bushman unknown, probably a derivation of the name tuiuva, regional nickname too complex.

Tataira - The Guarani tata (fire), anger (bee), bee fire, an interesting species, known in Brazil for his queimante acid deposited on the skin,

A person who bothers stir in its nest.

Healthful pequeninas.medem 4 X 3 mm, orange head and abdomen and thorax black, ocher smell. (Mustard) The tataira has a little bend in the posterior tarsus, which denotes rapine, so it is a harmful species to establish in meliponary, although some say that your honey is good.

The shifting cultivators to use home remedies without are problems.

Xope - the largest indigenous bee Irapuá, a subspecies of it, all gleaming black, with the tips of the wings white or light grey, the input is thick straw cardboard sounding horn. They nest in large trees Goiás, through the branches, round in shape and white colour. The interesting thing is that this species has excellent wax, great to roll candles and lanterns, the Indians and mestizos, use it to brighten the night's backcountry. A detail - these bees have good honey, caramel flavour but do not like noise and violently attack, curling hair and biting the skin of those who disturb. Farmers in this region used to bring these bees near the plantations of fruit trees, for them, afujentarem birds at harvest time.

Faria AC/2009

Vernacular Terms Glossary

Meliponídeos - scientific classification of all native bees in Brazil.
Beekeeping - beekeeping indigenous
Meliponidae - good honey bees, which have stunted the sting:
Native bees - wild bees of Brazil, whose names given by the Indians in the language, and for each species according to their behavior and experience. Ex:
Jati (nickname Jatahy) - yellow sweet honey bees (Tupy)
Irapuá - Honey Round (Guaná) Terenas-60
Jandaíra - big bee, hairy, much honey and thorns (Tupy)
Annatto - copper pot or in the ground (Tupy)
Manduri - brave, bold, courageous (Guaná)
Coeirupu - resident in the highlands (Guaná)
Guarupu - dweller in fertile land and good planer (Tupy)
Mandaçaia - bee fly, very gentle.
Pindamonhangaba - waters where fish have large daily. (Guaná)
Pindamonhangaba - factory hooks. (Tupy)

Table 4 " NATIVE SPECIES OF BEES Brazilian stingless "

Folk Name	**Popular Name**	**Scientific Name**
Abelha cachorro	Bee Dog	Trigona
Abelha do cupim	Bee termite	Aparatrigona
Abelha limão / Iratim	Bee lemon	Lestrimelitta
Abelha Raiz	Bee Root	S/N
Beiço	Lip	Melipona
Bieira	Port	Mourella
Boca de renda	Mouth income	Melipona
Boca de sapo	Mouth despot	Partamona
Borá	Type of Honey Bee	Tetragona quadrangula (Lepeletier
Bugia, Tujuba	Candle, Slave dance	Melipona
Bunda de Vaca	Ass Cow	Trigona
Caga fogo	Shits fire "ant"	Oxytrigona
Canudo	Straw	Scaptotrigona
Canudo	Straw	Scaptotrigona
Cupira preta	Black Cupira	Partamona
Feiticeira	Witch	Trigona
Guaraipo	Cunning Person	Melipona
Guaraipo amarela	Yellow guaraipo	Melipona
Guarupu	Fura (fluorescent)-earth	Melipona
Guaxupé	Guaxupé	Trigona
Guira	Musical Instrument	Geotrigona
Guiruçú/ Mel do chão	Guiruçú / Mel floor	Schwarziana
Irai	Irai	Nannotrigona
Irapuá	Irapuá	Trigona
Jandaira	Jandaira	Melipona
Jandaira amarela	Yellow Jandaira	melipona Crinita
Jatai	Jatai	tetragonisca angustula
Jataí Acreana	Jataí Acreana	Tetragonisca
jatai d terra	Jatai d earth	Paratrigona

Jataí da terra	Jataí Earth	Paratrigona
Jataí Negra	Black jataí	Scaura
Jupará	Jupará	Melipona
Lábios de morena	Lips brunette	Leurotrigona
Lambe olhos	Lambe eyes	Leurotrigona
Mandaçaia	Mandaçaia	Melipona
Mandaçaia	Mandaçaia	Melipona
Mandaçaia da terra	Mandaçaia Earth	Melipona
Mandaguari	Mandi	Scaptotrigona
Mandaguari amarela	Yellow Mandaguari	Scaptotrigona
Manduri	Mandi	Melipona
Manduri	Mandi	Melipona
Manduri amarela	Yellow Manduri	Melipona
Manduri do Matogrosso	Mandi's Matogrosso	Melipona
Marmelada amarela brava	Brave Yellow Marmalade	Frieseomelitta
Mirim	Mirim	Plebeia
Mirim guaçú	Mirim Guaçú	Plebeia
Mirim guaçú amarela	Mirim yellow	Plebeia
Mirim preguiça	Mirim laziness	Friesella
Moça branca ou mosquito	White girl or mosquitoes	Frieseomelita tricocerata
Moça preta	Black girl	Frieseomelitta
Mocinha Preta	Black Lassie	Friesiomelitta silvestre
Mombuca	Mombuca	Trigona
Mombucão	Mombucão	Cephalotrigona
Monoeca	Monoeca	Monoeca Xanthopyga
ninf	ninf	Paratrigona
Olho de Vidro	Glass Eye	Trigona
Pinto de Velho, Nariz de Anta	Pinto Old Nose Anta	Melipona

Rajada	Gust	Melipona
Saiqui	Saiqui	Plebeia
Sanharão	Sanharão	Trigona
Seminigra pernigra	Seminigra pernigra	Melipona
Tiúba do maranhão	Tiúba the maranhão	Melipona
Tubiba	Tubiba	Scaptotrigona
Tubuna	Tubuna	Scaptotrigona
Uruçu amarela preguiçosa	Uruçu lazy	Melipona
Uruçú Boca de Renda	URUCU Boca Income	Melipona
Uruçu boi	Uruçu Ox	Melipona
Uruçu verdadeira	Uruçu true	Melipona
Vorá	Vora	Tetragona clavipes

Web references and dictionaries

1. Dictionary.Com (http://dictionary.reference.com/)
2. The Free Dictionary by Farlex (http://www.thefreedictionary.com/)
3. Mnemonic Dictiionary (http://www.mnemonicdictionary.com/)
4. Lexic.us (http://www.lexic.us/)
5. Omnilexica *search all dictionaries at once!* (http://www.omnilexica.com)
6. Babylon Dictionary version 10.0.1(r18)
7. eol encyclopedia of life (http://eol.org/)
8. Discover life (http://www.discoverlife.org/)
9. naturalista (http://conabio.inaturalist.org/)
10. English-Word Informatiion (http://wordinfo.info/)
11. LatDict Latin dictionary anf grammar resources (http://www.latin-dictionary.net/)
12. Wiktionary, the free dictionary(http://en.wiktionary.org/wiki/Wiktionary:Main_Page)
13. BEE GENERA OF THE WORLD - *based on Michener, 2007.* Version 9-31-2007 (http://cache.ucr.edu/~heraty/beepage.html)
14. *Online Etymology Dictionary* **http://www.etymonline.com**
15. Zipcodezoo online http://zipcodezoo.com/default.asp

Works Cited

Kevan P & Imperatriz Fonseca VL (eds). (2002). *POLLINATING BEES.* Brasilia: Ministério do Meio Ambiente.

*#Koch, H. (2010). Combining morphology and DNA barcoding resolves the taxonomy of Western Malagasy Liotrigona Moure, 1961 (Hymenoptera: Apidae: Meliponini). *6MADA* (http://www.thefreelibrary.com/).

A. Pauly & N.Vereecken. (2013). *African Meliponinae.* Retrieved 29 April, 2014, from Atlas Hymenoptera: http://www.atlashymenoptera.net/page.asp?ID=121

A. R. Rech & M. L. Absy. (2011). Pollen storages in nests of bees of the genera Partamona, Scaura and Trigona (Hymenoptera, Apidae). *Revista Brasileira de Entomologia 55(3): setembro,* , 361–372,.

Barrackpore, S. (Decenber, 2001). *Barrackpore.* Retrieved 30 April, 2014, from http://www.barrackpore.gov.in/HTM/sdobkp_History_Recent.htm

BEARMAN, J. (22 February, 2011). *NATURALISTS' DEATHS BY UNNATURAL CAUSES.* Retrieved 21 April, 2014, from The Rumpus: http://therumpus.net/2011/02/naturalists-deaths-by-unnatural-causes/

Bishop Museum . (1983). J. LINSLEY GRESSITT (1914-1982). *INTERNATIONAL JOURNAL OF ENTOMOLOGY* .

Bohart, G. E. (1970). The Evolution of Parasitism Among Bees.

Boongird, M. &. (2004). A New Species of Trigona from Peninsular Thailand (Hymenoptera: Apidae: Meliponini).

Camargo, J. (2008). *Nest of Leurotrigona pusilla.* Retrieved 29 April, 2014, from Springer Images: http://www.springerimages.com/Images/LifeSciences/1-10.1007_978-1-4614-4960-7_2-9

Cebaehren. (29 January, 2012). *A New Interpretation of the Oldest Fossil Bee - Cretotrigona prisca.* (American Museum of Natural History) Retrieved 15 April, 2014, from Abejas, hormigas, avispas del Neotrópico: http://ordenhymenoptera.blogspot.com/2012/01/trigona-prisca

Claus Rasmussen & John S. Ascher. (2008). Heinrich Friese (1860–1948): Names proposed and notes on a pioneer melittologist. *Zootaxa* .

CLAUS RASMUSSEN et al. (2009). Curt Schrottky (1874–1937): South American entomology at the beginning of the 20th century. *Zootaxa* .

Cockerell, T. D. (1917). *New Social Bees.* Retrieved 28 April, 2014, from Psyche 24: http://psyche.entclub.org/24/24-120.html

Cockerell, T. D. (1917). New Social Bees. *Psyche 24 , 120* (August), 120-128.

Databank, B. E. (17 June, 2013). *Warwick Estevam Kerr.* Retrieved 21 April, 2014, from http://www.ebras.bio.br/entomol/entomol_desc.asp?code=0E1E3B078

Databank, B. E. (17 June, 2013). *Padre Jesus Santiago Moure*. Retrieved 21 April, 2014, from Entomologists in Brazil: http://www.ebras.bio.br/entomol/entomol_desc.asp?code=B0EEF21C5

Doug Yanega. (31 September, 2007). *BEE GENERA OF THE WORLD based on Michener, 2007*. Retrieved from http://cache.ucr.edu/~heraty/beepage.html

Eardley. (2004). - Taxonomic revision of the African stingless bees(Apoidea: Apidae: Apinae: Meliponini).

Eardley & Urban. (2010). Catalogue of Afrotropical bees. *2455*.

Encyclopidie, N. (18 March, 2014). *Encyclo*. Retrieved 28 April, 2014, from online Encyclopidie: http://www.encyclo.nl/

ENGEL, M. S. (2001). A MONOGRAPH OF THE BALTIC AMBER BEES AND EVOLUTION OF THE APOIDEA (HYMENOPTERA). *BULLETIN OF THE AMERICAN MUSEUM OF NATURAL HISTORY* .

Gazi, T. (2013). Pendlebury Collection. *B u l l e t i n o f t h e M u s e u m o f Z o o l o g y* , 5.

GBIF. (1 July, 2013). *GBIF Backbone Taxonomy*. Retrieved 6 May, 2014, from The Global Biodiversity Information Facility: http://www.gbif.org/species/4677363

Geni. (2014). *Harold Joseph Hockings (1856 - d.)*. Retrieved from Family Tree Geni: http://www.geni.com/people/Harold-Joseph-Hockings/6000000012184511355

Hutchison, R. H. (1921). *The Mulford Biological Exploration*. Retrieved from ENTOMOLOLirAL NEWS 245: http://archive.org/stream/cbarchive_49140_themulfordbiologicalexploratio1890/themulfordbiologicalexploratio1890_djvu.txt

ITIS. (5 March, 2014). *Bee Checklist*. Retrieved 21 April, 2014, from http://www.itis.gov/beechecklist.html

ITIS. (Sep , 2009). *Catalogue of Life: 2011 Annual Checklist*. Retrieved 29 April, 2014, from ITIS Bees Species 2000: http://www.catalogueoflife.org/annual-checklist/2011/details/species/id/6971938

Jadot, J.-M. (1968). Biographie Belge d'Outre Mer. *VI*.

Jarau & Barth. (2008). Stingless bees of the Golfo Dulce region, Costa Rica (Hymenoptera, Apidae, Apinae, Meliponini).

Jstor. (19 April , 2013). *Drescher, Friedrich Carl (1875-1957)*. Retrieved 20 April, 2014, from http://plants.jstor.org/person/bm000347660?s=t

Kelner-Pillault, S. (1969, 1970). Annales de la Société Entomologique de France 6. *Paleobiology Database* , 437-441.

Kerr W, Maule V. (1964). Geographic distribution of stingless bees and its implications. *Journal of the New York Entomological Society* , 72:2–18.

Krug, C. (2010). The community of bees (Hymenoptera - Apoidea) of two areas of biological interest and history in Santa Catarina. *University of São Paulo* .

Marianist. (2014). *Giovani Onore*. Retrieved 21 April, 2014, from http://www.marianist.com/articles/onore.htm

Michener, C. D. (2000). Bees Of The World. JHU Press.

Moure et al. (1958). Evolutionary problems among Meliponinae (Hymenoptera, Apoidea).

Moure J.S. & Camargo J.M.F. (1978). Fossil Stingless Bee From Copal (Hymenoptera: Apidae).

Moure's Bee Catalogue. (23 Jul, 2008). *Bee Catalogue*. (Universidade Federal do Paraná, UFPR, Centro de Referência em Informação Ambiental, CRIA) Retrieved 17 Jun, 2013, from Moure's Bee Catalogue: http://moure.cria.org.br/catalogue

NIEH, J. C. (Spring, 2000). *DAVID WARD ROUBIK* . Retrieved 21 April, 2014, from Scientific American Frontiers: http://www.pbs.org/safarchive/3_ask/archive/bio/81_jniehanddroubik_bio.html

Nilsson, A. N. (26 January , 2014). *Anders N. Nilsson*. Retrieved 21 April, 2014, from International Homepage: http://www2.emg.umu.se/projects/biginst/andersn/index.htm

Nogueira-Neto, P. (1997). Life Creating Bees.

Rasmussen. (2008). Catalog of the Indo-Malayan/Australasian stingless bees (Hymenoptera: Apidae: Meliponini).

Rasmussen, C. (2012). Biography Joseph Vachal. *3442*.

Rasmussen, C. (2013). Personal communication.

REHN, J. A. (1989). *The Stanford Expedition to Brazil, 1911*. Retrieved from Trans. Am. Ent. Soc, Vol. XLIII.: http://archive.org/stream/jstor-25076965/25076965_djvu.txt

Ruggiero, M. (3 October, 2008). *(ITIS) Bee Checklist,*. (ITIS, United States.) Retrieved from Integrated Taxonomic Information System: http://www.itis.gov/beechecklist.html

Schwarz. (1938).

Schwarz, H. F. (1939). *The Indo-Malayan Species of Trigona* .

Senckenberg Biographies. (11 December, 2013). *World of Biodiversity*, 6. (Senckenberg) Retrieved 20 April, 2014, from Biographies of entomologists of the world: http://sdei.senckenberg.de/biographies/

Smith, G., & Walker, C. a. (2011). What's in a name: an update on epithets used in Aloe L. (Asphodelaceae). *Bradleya, 29 pp.* , 179–181. .

Torres, M. A. (July, 2005). *Native Bees ASF*. Retrieved 28 April, 2014, from Apisguia: http://www.apisguia.com.br/?pagina=principal

UFPR. (2012). *Danuncia Urban*. Retrieved 20 April, 2014, from http://www.ufpr.br/portalufpr/noticias/cnpq-destaca-professoras-da-ufpr-entre-pioneiras-da-ciencia-no-brasil/

UNESP. (14 April, 2014). *Rectory Portal*. (UNESP) Retrieved 20 April, 2014, from http://www.unesp.br/portal#!/eng

Virtual Museum. (1998-99). *Giovanni Gribodo*. Retrieved 20 April, 2014, from Politecnico di Torino.: http://areeweb.polito.it/strutture/cemed/museovirtuale/english/storia/2-02/2-2-03/2-2-0331a.htm

WERNER., A. (September , 1890). *The African Pygmies*. Retrieved 28 April, 2014, from Popular Science Monthly | Volume 37 | : http://en.wikisource.org/wiki/Popular_Science_Monthly/Volume_37/September_1890/The_African_Pygmies

Wikia. (n.d.). *Biologie der brasilianischen*. Retrieved 5 May, 2014, from Apicultura Wikia: http://apicultura.wikia.com/wiki/Letra_I1

Wikia. (2014). *Herbert Ferlando Schwarz*. Retrieved 21 April, 2014, from http://apicultura.wikia.com/wiki/Herbert_Ferlando_Schwarz

Wikia. (n.d.). *Jesús Santiago Moure*. Retrieved 21 April, 2014, from http://apicultura.wikia.com/wiki/Jes%C3%BAs_Santiago_Moure

Wikia. (2007). *João M. F. de Camargo*. (Apicultura Wiki) Retrieved 20 April, 2014, from http://apicultura.wikia.com/wiki/Jo%C3%A3o_M._F._de_Camargo

Wikipedia. (7 February , 2014). *Adolpho Ducke*. Retrieved 20 April, 2014, from http://en.wikipedia.org/wiki/Adolpho_Ducke

Wikipedia. (29 December , 2013). *Alexander W. von Götte* . Retrieved 20 April, 2014, from http://en.wikipedia.org/wiki/Alexander_W._von_G%C3%B6tte

Wikipedia. (12 April, 2014). *Alexandre Rodrigues Ferreira* . Retrieved 20 April, 2014, from http://en.wikipedia.org/wiki/Alexandre_Rodrigues_Ferreira

Wikipedia. (3 December, 2013). *Alfred E. Emerson*. Retrieved 20 April, 2014, from http://en.wikipedia.org/wiki/Alfred_Edwards_Emerson

Wikipedia. (9 April , 2014). *Alfred_Russel_Wallace*. Retrieved 21 April, 2014, from http://en.wikipedia.org/wiki/Alfred_Russel_Wallace

Wikipedia. (11 July , 2012). *Alvaro Wille*. Retrieved 21 April, 2014, from https://es.wikipedia.org/wiki/%C3%81lvaro_Wille_Trejos

Wikipedia. (12 December , 2013). *Angophora*. Retrieved 28 April, 2014, from http://en.wikipedia.org/wiki/Angophora

Wikipedia. (25 March, 2014). *Atahualpa*. ("Atahualpa". Appletons' Cyclopædia of American Biography. 1900.) Retrieved 20 April, 2014, from http://en.wikipedia.org/wiki/Atahualpa

Wikipedia. (8 April , 2014). *Cândido Rondon*. Retrieved 21 April, 2014, from http://en.wikipedia.org/wiki/Marechal_Rondon

Wikipedia. (10 November , 2013). *Cecil Boden Kloss* . Retrieved 20 April, 2014, from http://en.wikipedia.org/wiki/Cecil_Boden_Kloss

Wikipedia. (19 April , 2014). *Ceratitis citriperda*. Retrieved 28 April , 2014, from http://en.wikipedia.org/wiki/Ceratitis_citriperda

Wikipedia. (5 February , 2014). *Charles Duncan Michener*. Retrieved 21 April, 2014, from http://en.wikipedia.org/wiki/Charles_Duncan_Michener

Wikipedia. (21 March , 2013). *Charles Henry Tyler Townsend*. Retrieved 21 April, 2014, from http://en.wikipedia.org/wiki/Charles_Henry_Tyler_Townsend

Wikipedia. (29 December , 2013). *Charles Thomas Bingham* . Retrieved 20 April, 2014, from http://en.wikipedia.org/wiki/Charles_Thomas_Bingham

Wikipedia. (25 December, 2013). *Correndera Pipit*. Retrieved 28 April, 2014, from http://en.wikipedia.org/wiki/Anthus_correndera

Wikipedia. (28 April , 2013). *Crodowaldo Pavan*. Retrieved 21 April, 2014, from http://en.wikipedia.org/wiki/Crodowaldo_Pavan

Wikipedia. (23 September , 2013). *Eduard Heinrich Graeffe*. Retrieved 20 April, 2014, from http://en.wikipedia.org/wiki/Eduard_Heinrich_Graeffe

Wikipedia. (16 January , 2014). *Émil Goeldi*. Retrieved 20 April, 2014, from http://en.wikipedia.org/wiki/%C3%89mil_August_Goeldi

Wikipedia. (14 October, 2013). *Emmanuel Bertrand-Bocandé* . Retrieved 20 April, 2014, from http://en.wikipedia.org/wiki/Emmanuel_Bertrand-Bocand%C3%A9

Wikipedia. (8 April , 2014). *Ernst Mayr*. Retrieved 21 April, 2014, from http://en.wikipedia.org/wiki/Ernst_Mayr

Wikipedia. (23 March , 2014). *Étienne Geoffroy Saint-Hilaire*. Retrieved 21 April, 2014, from http://en.wikipedia.org/wiki/%C3%89tienne_Geoffroy_Saint-Hilaire

Wikipedia. (30 April , 2013). *Filippo Silvestri*. Retrieved 21 April, 2014, from http://en.wikipedia.org/wiki/Filippo_Silvestri

Wikipedia. (20 July, 2013). *Frederick Smith*. Retrieved 21 April, 2014, from http://en.wikipedia.org/wiki/Frederick_Smith_(entomologist)

Wikipedia. (21 February, 2014). *Frederick William Beechey*. Retrieved 20 April, 2014, from http://en.wikipedia.org/wiki/Frederick_William_Beechey

Wikipedia. (29 December , 2013). *Friedrich Moritz Brauer* . Retrieved 20 April, 2014, from http://en.wikipedia.org/wiki/Friedrich_Moritz_Brauer

Wikipedia. (14 January , 2014). *George Gaylord Simpson*. Retrieved 21 April, 2014, from http://en.wikipedia.org/wiki/George_Gaylord_Simpson

Wikipedia. (19 August , 2013). *Gnaphosidae*. Retrieved 28 April, 2014, from http://en.wikipedia.org/wiki/Gnaphosidae

Wikipedia. (17 April, 2014). *Gregor Mendel*. Retrieved 21 April, 2014, from http://en.wikipedia.org/wiki/Gregor_Mendel

Wikipedia. (28 December , 2013). *Henry Edward Crampton* . Retrieved 20 April, 2014, from http://en.wikipedia.org/wiki/Henry_Edward_Crampton

Wikipedia. (1 March, 2014). *Henry Walter Bates*. Retrieved 20 April, 2014, from http://en.wikipedia.org/wiki/Henry_Walter_Bates

Wikipedia. (6 March , 2014). *Hermann Muller*. Retrieved 21 April, 2014, from http://en.wikipedia.org/wiki/Heinrich_Ludwig_Hermann_M%C3%BCller

Wikipedia. (27 October , 2013). *Johannes Hildebrandt*. Retrieved 20 April, 2014, from http://en.wikipedia.org/wiki/Johannes_Hildebrandt

Wikipedia. (15 April , 2014). *John Casper Branner* . Retrieved 20 April, 2014, from http://en.wikipedia.org/wiki/John_Casper_Branner

Wikipedia. (17 February , 2014). *Karl Wilhelm von Dalla Torre* . Retrieved 20 April, 2014, from http://en.wikipedia.org/wiki/Karl_Wilhelm_von_Dalla_Torre

Wikipedia. (25 December , 2013). *Lineated Woodpecker*. Retrieved 28 April, 2014, from http://en.wikipedia.org/wiki/Dryocopus_lineatus

Wikipedia. (31 December, 2013). *Maurits Lieftinck*. Retrieved 21 April, 2014, from http://nl.wikipedia.org/wiki/Maurits_Lieftinck

Wikipedia. (15 June , 2013). *Melbourne Armstrong Carriker* . Retrieved 20 April, 2014, from http://en.wikipedia.org/wiki/Melbourne_Armstrong_Carriker

Wikipedia. (19 April , 2014). *Mulatto*. Retrieved 28 April, 2014, from http://en.wikipedia.org/wiki/Mulatto

Wikipedia. (10 April , 2014). *Otto Staudinger*. Retrieved 21 April, 2014, from http://en.wikipedia.org/wiki/Otto_Staudinger

Wikipedia. (6 December, 2013). *Owain Richards*. Retrieved 21 April, 2014, from http://en.wikipedia.org/wiki/Owain_Richards

Wikipedia. (28 January , 2014). *Prosopis*. Retrieved 28 April, 2014, from http://en.wikipedia.org/wiki/Prosopis

Wikipedia. (26 April , 2014). *Quagga*. Retrieved 28 April , 2014, from http://en.wikipedia.org/wiki/Equus_quagga_quagga

Wikipedia. (18 March , 2014). *Red Algae*. Retrieved 29 April, 2014, from http://en.wikipedia.org/wiki/Rhodophyta

Wikipedia. (20 April , 2014). *Red Fox*. Retrieved 28 April , 2014, from http://en.wikipedia.org/wiki/Vulpes_vulpes

Wikipedia. (16 August , 2013). *Sarothrura*. Retrieved 28 April, 2014, from http://en.wikipedia.org/wiki/Sarothrura

Wikipedia. (30 December , 2013). *Scale Insects*. Retrieved 28 April, 2014, from http://en.wikipedia.org/wiki/Coccid

Wikipedia. (26 February , 2013). *Scutellum*. Retrieved 29 April, 2014, from http://en.wikipedia.org/wiki/Scutellum_%28insect_anatomy%29

Wikipedia. (11 October, 2013). *Sheffield Airey Neave*. Retrieved 21 April, 2014, from http://en.wikipedia.org/wiki/Sheffield_Airey_Neave

Wikipedia. (12 February , 2014). *Shonen Matsumura*. Retrieved 21 April, 2014, from http://en.wikipedia.org/wiki/Shonen_Matsumura

Wikipedia. (12 March , 2014). *Solanum*. Retrieved 29 April, 2014, from http://en.wikipedia.org/wiki/Solanum

Wikipedia. (26 December , 2013). *Theodor Peckolt*. Retrieved 21 April, 2014, from http://en.wikipedia.org/wiki/Theodor_Peckolt

Wikipedia. (28 December , 2013). *Theodore Dru Alison Cockerell* . Retrieved 20 April, 2014, from http://en.wikipedia.org/wiki/Theodore_Dru_Alison_Cockerell

Wikipedia. (6 April , 2014). *Theodosius Dobzhansky*. Retrieved 20 April, 2014, from http://en.wikipedia.org/wiki/Theodosius_Dobzhansky

Wikipedia. (5 February , 2014). *Topi*. Retrieved 29 Aptil, 2014, from http://en.wikipedia.org/wiki/Topi

Wikipedia. (31 December , 2013). *Triplaris*. Retrieved 28 April, 2014, from http://en.wikipedia.org/wiki/Triplaris

Wikipedia. (3 February , 2014). *Vittorio Bottego* . Retrieved 20 April, 2014, from http://en.wikipedia.org/wiki/Vittorio_Bottego

Wikipedia. (18 February , 2014). *West Papua*. Retrieved 21 Aptil, 2014, from http://en.wikipedia.org/wiki/West_Papua_(province)

Wikipedia. (2 April , 2014). *William Beebe*. Retrieved 20 April, 2014, from http://en.wikipedia.org/wiki/Charles_William_Beebe

Wikipedia. (8 March , 2014). *William M. Mann*. Retrieved 21 April, 2014, from http://en.wikipedia.org/wiki/William_M._Mann

Wikipedia. (28 April , 2014). *Zebra*. Retrieved 29 April , 2014, from http://en.wikipedia.org/wiki/Zebra

Wikipedia. (23 March , 2013). *Zexmenia*. Retrieved 29 April, 2014, from http://en.wikipedia.org/wiki/Zexmenia

Zucchi, R. (23 August, 2009). *Ronaldo Zucchi*. Retrieved 21 April, 2014, from Curriculo Lattes: http://lattes.cnpq.br/8652192235566383

BIBLIOGRAPHY:

ASCHER & PICKERING, 2012.

Discover Life's bee species guide and world checklist,

BARNI G.S., et al.

Moretto Mitochondrial genome differences between the stingless bees *Melipona rufiventris* and *Melipona mondury* (Apidae: Meliponini)

BROWN R. W.1954

"Composition of scientific words; a manual of methods and a lexicon of materials for the practice of logotechnics."

BÄNZIGER et al.

The Remarkable Nest Entrance of Tear Drinking *Pariotrigona klossi* and Other Stingless Bees Nesting in Limestone Cavities (Hymenoptera: Apidae) Author(s) :

CHINH T. X., et al.

Nest and Colony Characteristics of Three Stingless Bee Species in Vietnam with the First Description of the Nest of *Lisotrigona carpenteri* (Hymenoptera: Apidae: Meliponini)

COSTA M. A. et al.

Molecular phylogeny of the stingless bees (Apidae, Apinae, Meliponini) inferred from mitochondrial 16S rDNA sequences

EARDLEY C. D., et al. 2009

Bee conservation in Sub-Saharan Africa and Madagascar: diversity, status and threats

EARDLEY C D

Taxonomic revision of the African stingless bees(Apoidea: Apidae: Apinae: Meliponini)

ENGEL M.S. 2001

&^ A Monograph Of The Baltic Amber Bees And Evolution Of The Apoidea (Hymenoptera)

ENGEL, M. S.

A New Interpretation of the Oldest Fossil Bee (Hymenoptera: Apidae)

ENGEL 2000

A Review of the Indo-Malayan Meliponine Genus *Lisotrigona*, with Two New Species (Hymenoptera: Apidae)

FRANCK P., et al.

Nest architecture and genetic differentiation in a species complex of Australian stingless bees

GONZALEZ V.H. & GRISWOLD T. L.

Two new species of *Paratrigona* and the male of *Paratrigona ornaticeps* (Hymenoptera, Apidae)

GONZALEZ V.H. & GRISWOLD T. L.

Two new species of *Paratrigona* and the male of *Paratrigona ornaticeps* (Hymenoptera, Apidae)

INOUE, et al. 1996

Individual Variations in Worker Polyethism of the Sumatran Stingless Bee, *Trigona (Tetragonula) minangkabau* (Apidae, Meliponae)

JAVIER J. G. et al.

Acereto Notes on Nest Habits and Host Range of Cleptobiotic Lestrimelitta nitkib (Ayala 1999)(Hymenoptera: Meliponini) From The Yucatan Peninsula.

JAVIER J. G. et al.

Genetic and phenotypic differentiation in endemic Scaptotrigona hellwegeri (Apidae: Meliponini): implications for the conservation of stingless bee populations in contrasting environments

JOBIRAJ & NARENDRAN 2004

A revised key to the world species *Lisotrigona* Moure (Hymenopter : Apoidea : Apidae) with description of a new species from India

JARAU & BARTH 2008

Stingless bees of the Golfo Dulce region, Costa Rica (Hymenoptera, Apidae, Apinae, Meliponini)

KRUG C. 2010

** A comunidade de abelhas (Hymenoptera – Apoidea) de duas areas de interesse biológico e histórico em Santa Catarina

KOCH H. 2010

Combining morphology and DNA barcoding resolves the taxonomy of Western Malagasy *Liotrigona* Moure, 1961 (Hymenoptera: Apidae: Meliponini)

K O C H H e t a l .

Honeydew collecting in Malagasy stingless bees (Hymenoptera: Apidae:Meliponini)and observations on competition with invasive ants

MICHENER C. D. & BOONGIRD S.

A New Species of *Trigona* from Peninsular Thailand (Hymenoptera: Apidae: Meliponini) Author(s):

MOURE J. S. & CAMARGO J. M. F. 1978

&& Fossil Stingless Bee From Copal (Hymenoptera: Apidae)

MOURE J. S. & CAMARGO J. M. F.

Melipona (michmelia) capixaba, uma nova espécie de melLponinae (Hymenoptera, Apidae) do sudeste do brasil

MOURE, J. S. 1961.

A Preliminary Supra-specific Classification of the Old World Meliponine Bees (Hymenoptera, Apoidea). Studia Entomologica, vol. 4, fasc. 1-4. 181-242

NJOYA M. T. M. 2009

The High Agricultural Faculty of the Rheinische Friedrich-Wilhelms-University presented at Bonn on 04 November 2009

Diversity of Stingless Bees in Bamenda Afromontane Forests – Cameroon:

Nest architecture, Behaviour and Labour calendar –

OLIVEIRA et al. 2011

The identity of the Neotropical stingless bee Frieseomelitta meadewaldoi (Cockerell, 1915) (Hymenoptera, Apidae) ZooKeys 111: 19–31 (2011)

RASMUSSEN C.

CATALOG OF the Indo-Malayan/Australasian stingless bees (Hymenoptera: Apidae: Meliponini)

RASMUSSEN & CAMERON 2009/2010

Global stingless bee phylogeny supports ancient divergence, vicariance, and long distance dispersal

ROUBIK D. W., et al.

New stingless bee genus endemic to Central American cloudforests: phylogenetic and biogeographic implications (Hymenoptera: Apidae: Meliponini)

SAKAGAMI & YAMANE 1984

Notes on Tazonomy and Nest Architecture of the Taiwanese Stingless Bee *Trigona (Lepidotrigona) ventralis hoozana*

SAKAGAMI 1978

Tetragonula Stingless Bees of the Continental Asia and Sri Lanka (Hymenoptera, Apidae)

WERNER A. 1890

~* Popular Science Monthly Volume 37 September 1890 (1890) The African Pygmies By A. Werner http://en.wikisource.org/wiki/Popular_Science_Monthly/Volume_37/September_1890/The_African_Pygmies

WINSTON J, E. 1999

D e s c r i b i n g S p e c i e s - **Practical Taxonomic Procedure for Biologists**

Melittosphex burmensis is the oldest-known species of bee. The species was discovered as an amber inclusion in the year 2006 by George Poinar, Jr., a zoologist at Oregon State University. The amber was found in a mine in the Hukawng Valley of northern Myanmar and is believed to date from the Cretaceous Period, 100 million years ago.

M. burmensis is approximately one-fifth the size of the extant honeybee, at about 3 millimetres long. It is not related to any existing bee family. *M. burmensis* has some anatomical features similar to those of flesh-eating wasps, including the shape of its hind legs, but also some features of pollen-collecting bees, such as branched hairs on the body. The bee's head is heart-shaped.

The sample discovered is thought to be 100 million years old, 40 million years older than the next-oldest known bee species. The discovery of a bee from the Cretaceous Period with certain pollen-collecting features may help to explain the rapid expansion of flowering plants at that time in earth's history.

Font colour key (Note: Entries not included here are in the General Index)

Black italics font Caps = Meliponine Genus Taxon

Dark Blue italics font Caps = Other Related Genus

Red, accent 2 (25%), italics font = Meliponine Species Patronyms & Matronyms

Dark Blue italics font = Ethnicity & Locality Epithets for Meliponine Species Taxa

Olive Green accent 3(50%) italics font = General Epithets for Meliponine Species Taxa

Orange, accent 6 (50%) italics font= Meliponine Anatomy and color attribute epithets.

Purple, accent 4 (25%) italics font= Meliponine Taxa with Flora & Fauna Relationship

Genus and Species Index (Scientific Name)

General Index

Inside Back Cover

The thin body profile of Tetragonisca angustula By: Walter Cecoto, Formosa, Argentina via Maria Rosa Bernasconi

Scutella of T. iridipennis with visible hair bands

The reddish Antennae of Tetragonula ruficornis

Photos by: Claus Rasmussen

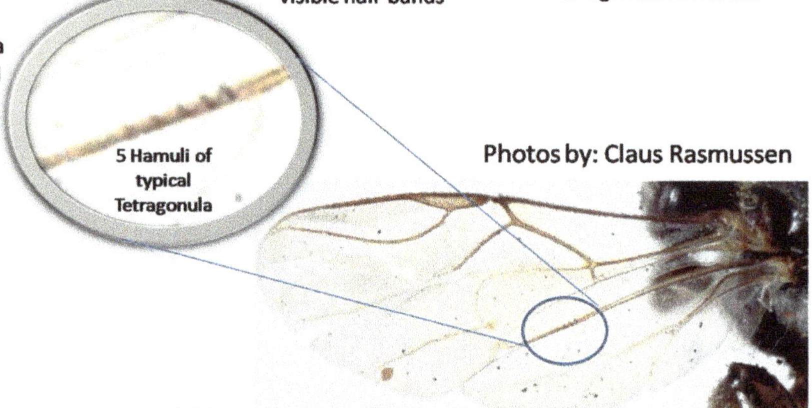

Iridescent wings of Tetragonula iridipennis

Did you know?

"The world's smallest bee is a dwarf species called ***Trigona minima***; it measures about 0.08-in long and is one of the species of stingless bees. At the other end of the scale, the largest bee in the world is an Indonesian leaf cutter bee called *Megachile pluto*, which attains 1.5-in in length. The most common species of bee worldwide is the European honeybee, our old friend *Apis Mellifera*, but in North America the most common kind is a type of halictid or 'sweat bee' (*Augochloropsis* Species), whose common name comes from the fact that they often land on people's skin to lick up the salt from their perspiration. This can be an alarming experience, since they often resemble wasps in appearance. Sweat bees also have an unusual way of obtaining pollen, known as buzz pollination, in which the bee grasps the anther in its jaws and vibrates its wings, causing the pollen to be dislodged onto its body." (Hall p.69-70, 2010)

9781913811044